John Wood

On Rupture, Inguinal, Crural, and Numerical

The anatomy, pathology, diagnosis, cause and prevention. With new methods of effecting a radical and permanent cure; embodying the jacksonian prize essay of the Royal College of Surgeons, 1861

John Wood

On Rupture, Inguinal, Crural, and Numerical

The anatomy, pathology, diagnosis, cause and prevention. With new methods of effecting a radical and permanent cure; embodying the jacksonian prize essay of the Royal College of Surgeons, 1861

ISBN/EAN: 9783337070458

Printed in Europe, USA, Canada, Australia, Japan

Cover: Foto ©berggeist007 / pixelio.de

More available books at **www.hansebooks.com**

ON RUPTURE,

INGUINAL, CRURAL, AND UMBILICAL.

PREPARING FOR PUBLICATION, BY THE SAME AUTHOR.

THE

SURGICAL ANATOMY OF THE PELVIS AND PERINEUM,

In the Infant, Young, and Adult Male and Female;

CONSTITUTING

A GUIDE TO OPERATIONS UPON THE URETHRA, BLADDER, RECTUM, AND OTHER PELVIC VISCERA,

And to Diagnosis in some of their Diseases.

ON RUPTURE,

INGUINAL, CRURAL, AND UMBILICAL;

THE ANATOMY, PATHOLOGY, DIAGNOSIS, CAUSE AND PREVENTION;

WITH

New Methods of effecting a Radical and Permanent Cure;

EMBODYING

THE JACKSONIAN PRIZE ESSAY OF THE ROYAL COLLEGE OF SURGEONS, LONDON, FOR 1861.

With Numerous Illustrations.

BY

JOHN WOOD, F.R.C.S., Eng. (Exam.)

DEMONSTRATOR OF ANATOMY AT KING'S COLLEGE, LONDON;
ASSISTANT-SURGEON TO KING'S COLLEGE HOSPITAL;
SURGEON TO THE LINCOLN'S INN DISPENSARY.

LONDON:
JOHN W. DAVIES, 54, PRINCES STREET,
LEICESTER SQUARE.

EDINBURGH: MACLACHLAN AND CO.   DUBLIN: FANNIN AND CO.

MDCCCLXIII.

The right of Translation is reserved.

PREFACE.

BEFORE giving to the profession the results of some years' careful study and experience of the various forms of rupture, and of the possibility of curing permanently a deformity so common and so disabling, I have fully felt the importance of a sufficient lapse of time, or duration, as a test of the reality of the cure.

Chiefly for this reason, and partly to enable me to make such modifications of my original method as might simplify and render more effective the steps of the operation, I have hitherto deferred this publication, preferring rather to answer privately, as far as I could possibly do so, the numerous applications I have had, both from the provinces and from abroad, for descriptions of my latest plans.

Having been honoured by the award of the Jacksonian prize for 1861, of the Royal College of Surgeons, London, for my Essay on "The best Method of effecting a Radical Cure of Inguinal Hernia," I feel it now incumbent upon me to embody the contents of that essay in the present publication. Much of the following pages has been added to the original essay.

The anatomical descriptions have been extended, with a view to the better comprehension of the parts concerned in an operation difficult to be understood by those who have not an every-day familiarity with the structures involved.

The section on the use of Trusses has been also amplified; the description of a new truss, without actual inspection of the specimens which accompanied the essay, requiring more minuteness, and involving a consideration of the action of trusses generally, in preventing or promoting a radical cure, both with and without an operation.

That part of the work which treats of Femoral and Umbilical Rupture has been added entire to the original essay. The former subject enables me to make public an operation which I have been in the habit for some years of showing on the dead subject to the pupils of King's College, but which, for reasons given in the text, I have not yet performed upon the living body; and also a description of a truss which I have employed with a view of promoting, in small cases, a radical cure by pressure at the most effective point. Under the head of Umbilical Rupture I have given for the first time to the public, an operation which I have performed with very good results, both as regards the absence of serious symptoms in the after progress of the cases, and their ultimate result. Upon the principles laid down in the text, I have devised a truss, or compress, for umbilical protrusion, which is herein figured and published for the first time. It is probably right to state here, that the drawings from which the illustrations are taken are all original, and made by the author from actual dissections, operations, and instruments.

It is not, in the present treatise, contemplated to include the whole of the points connected with the diagnosis and treatment of the various kinds of hernia, nor to attempt a substitute for the valuable and compendious works of Sir A. Cooper and Mr. Lawrence upon this subject, but simply to take such points as it is important to consider in effecting a radical cure.

Centuries of experience of the futility of proposed plans for curing rupture have had the effect of producing an excusable hesitation and doubt in the profession as to the reality and permanence of the cures produced.

PREFACE.

The liberality of my colleagues, the senior surgeons of King's College Hospital, has enabled me publicly to test my various methods in most of the cases given in the Appendix; and from time to time to exhibit in the theatre of that institution the condition of many of the patients, at periods of time after the operation varying from six months to five years. In many cases, the additional test of heavy labour has been applied for more than a year, without truss or any support whatever. Six of the cases given in the Appendix to this volume were publicly exhibited at the meeting of the Medico-Chirurgical Society, in February, 1860. In all the cases I have been at much pains to ascertain, as far as circumstances would allow, the condition of the patients up to the present time.

To numerous friends, whose generous sympathy for the advancement of surgical science has led them to second my efforts in a direction considered by many as hopeless, and in which, but for such support, I should have found it impossible to proceed, my best thanks are here sincerely tendered.

4, MONTAGUE STREET, RUSSELL SQUARE, W.C.
April, 1863.

ANALYTICAL TABLE OF CONTENTS.

INTRODUCTION.

	PAGE
Demand for a radical cure of hernia	2
Supposed dangers from peritonitis	3
Inconvenience and risks from strangulation and from an operation compared	4
Risk compared with that of similar operations	5
Degree of safety and probabilities of success	6
Success of other modern efforts for a radical cure	7
Proper choice of cases	8
Superior eligibility of inguinal hernia	9

PART I.

ON INGUINAL HERNIA.

SECTION I.—ANATOMY OF THE PARTS.

Superficial dissection—fasciæ, &c.	10
,, ,, vessels	12
General arrangement of the lower abdominal wall and groins	13
Internal abdominal ring and fasciæ	15
Position and length of inguinal canal	15
External abdominal rings, its pillars and fasciæ	16
Anterior wall—cremaster muscle and fascia	17
Posterior wall—transversalis fascia	19
,, ,, conjoined tendon	20
,, ,, triangular aponeurosis	21
Superior boundary of the canal	21
Inferior boundary, rectus muscle	22

	PAGE
Peritoneum—superior false ligament of bladder	23
Subperitoneal fascia, external iliac artery	24
Deep circumflex iliac	25
,, epigastric and branches	26
Triangle of Hesselbach, vas deferens, spermatic cord and vessels	27
Composition, course, and coverings of cord	28

SECTION II.—CAUSES AND PATHOLOGY OF INGUINAL HERNIA.

Theory of elongated mesentery—objections	30
Formation of sac	33
Failure of the tendinous walls and openings	35
Varieties of inguinal hernia	36
Oblique hernia and coverings	37
Direct hernia and coverings	39
Influence of the rectus muscle	40
Bubonocele—relations to spermatic cord	41
Neck of the sac—influence on the abdominal rings	42
Varieties in the sac coverings	44
Different points of contraction in the canal	46
Differences between recent and old formations	47
Varieties in shape and outline of the sac	48
General characteristics of hernial subjects	51
Congenital hernia and its formation	58
Causes—canal of Nuck	59
Infantile hernia—mode of formation	60
Direct hernia in the child	61

SECTION III.—DIAGNOSIS OF INGUINAL HERNIA.

Physical signs—cough impulse	62
Modifications of impulse	64
Mode of formation and progress	65
Effect of manipulation or *taxis*	66
Position of testis, imperfect descent of testis	67
Fatty tumours, cysts	68
Hydrocele of the cord	69
Glandular tumours	70
Abscesses, hydrocele of tunica vaginalis	71
Enlarged testis, varicocele	73

Section IV.—History of the Radical Cure of Inguinal Hernia.

	PAGE
Ancient methods—ointments, plasters, actual cautery, caustics	76
Excision of sac, ligature, royal stitch, punctum aureum	77
Older methods—operations of Schmucker and Langenbeck	78
Richter and L'Estrange's methods by pressure	79
Modern operations. 1st class—Schuh's, Riggs', Belmas', Velpeau's, and Pancoast's	80
Bonnet's method. 2nd class—Gerdy's and Signoroni's methods	81
Wurtzer's method—its advantages	82
Objections—causes of failure	84

Section V.—Principles of the Author's Operations.

Adaptations of anatomical peculiarities	88
Differences from former operations	90
Variations in the means employed, viz.—thread, wire, and pins	92
Use of external compress, advantages of metallic sutures	93
Main principles of the operations	94

Section VI.—Operation by Thread and Compress.

Description of the instruments employed	95
Description of the steps of the operation	98
Arrangement of the apparatus in reference to the anatomy of the parts	104
Dressings, after-course, and treatment	106

Section VII.—Operation by Wire as usually practised by the Author.

Instruments used, choice of wire, &c.	107
Description of the steps of the operation	109
Dressings, after-course, and treatment	114

Section VIII.—Variations of the Wire Operation.

Variations with single suture	119
,, ,, double cross suture	122
Another variation with double suture	124
Variation with transverse sutures	127

Section IX.—Operation by Rectangular Pins.

	PAGE
Description of rectangular pins	130
Longitudinal application of pins	132
Method of dressing	134
Transverse application of pins	136
After-treatment and progress of cases	137

Section X.—Modus Operandi of Operations.

Effect upon the tendinous hernial canal	140
Illustrative cases—effect upon the sac	142
Effect upon the invaginated scrotal fascia	144
Effect upon the external ring	145
Direct action upon the tissues	146

Section XI.—Causes of Failure and Danger.

Shape and application of instruments	147
Protection to the large vessels and bowel	149
Failure in securing the conjoined tendon and remedy	151
Malposition of the sutures	152
Relation of the sutures to the spermatic cord	153
Difficulties of direct cases	154
Importance of sufficient separation of scrotal fascia	155
Importance of good health in the patient	156
Degree of compression to be employed	157
Choice of operation for youths and children—small cases	158
Best operation for adults and large cases	159

Section XII.—Summary of the Cases.

Cause of death in the single fatal case, average duration of the cases	160
Number and methods of operating in the 60 cases given in the Appendix	161
Number and causes of the failures in the operation by thread and compress	162
Number and causes of the failures in the operation by wire	162
Number and causes of the failures in the operation by pins	163
After-condition of the ruptures in unsuccessful cases	164

CONTENTS. xiii

	PAGE
Numbers and condition of the doubtful cases, cases of repetition of the operation	165
Consequences to the testis in three cases	166
Total result and percentage	167

SECTION XIII.—USE OF TRUSSES IN INGUINAL HERNIA.

Injurious effects of convex truss-pads	168
Effect of efficient truss pressure	169
Description of horse-shoe boxwood truss-pad for oblique hernia	170
„ „ lever-spring pad	173
„ „ H-spring pad	174
Method of application and mechanical effects of the foregoing	175
Description of the ovoid ring pad for direct hernia	176
„ „ lever-spring ring pad	177
„ „ double pad and spring truss for large cases	178

PART II.

FEMORAL OR CRURAL HERNIA.

SECTION XIV.—ANATOMY OF THE PARTS.

Scarpa's triangle, superficial fasciæ	183
Superficial vessels and nerves, &c.	184
Arrangement of deep fascia	186
Description of the saphenous opening and cribriform fascia	187
Falciform border—Heys' ligament	188
Infundibular crural sheath—deep arch	189
Lymphatic compartment—crural ring	190
Septum crurale—crural canal	191
Relations to important vessels	192
Important and dangerous irregularities	194
Abdominal relations of crural ring	196
Irregularities of obturator artery	197

SECTION XV.—CAUSES AND PATHOLOGY OF CRURAL HERNIA.

Causes of its prevalence in the female	200
Abdominal aspect and axis of crural ring	201

xiv   CONTENTS.

	PAGE
First direction of the rupture—relation to the femoral vein	202
Final direction of the rupture and its causes	203
Alteration of the axis and its consequences	204
Coverings of crural hernia	205
Alterations of structure in old cases	206
Varieties and their causes	207
Conditions favourable for a radical cure	208

Section XVI.—Diagnosis of Crural Hernia.

Position, shape, and other signs distinguishing from the inguinal variety	210
Effect of obesity, direction of cough impulse	212
Enlarged glands	213
Fatty tumours	214
Cysts	215
Chronic abscess, &c.	216

Section XVII.—Operation for the Radical Cure.

Operations by Dr. Jameson and Mr. Davies	217
The author's operation by wire	218
Effect of the operation—necessary precautions	221
Risks of, and proper subjects for operation	222
Radical cure after the operation for strangulation	223

Section XVIII.—Radical Cure by Truss Pressure.

Importance of a proper application of the pressure	224
Description of the elliptical truss pad	226
,, ,, ,, lever-spring pad	227

Section XIX.—Prevention of Inguinal and Crural Hernia.

Classes most affected, predisposing habits	229
Signs of a tendency to rupture	230
Description of a preventive bandage for weak groins	231

Section XX.—Treatment of Irreducible Hernia.

Truss pressure in omental hernia	234
Operation for radical cure in ditto	236

PART III.

UMBILICAL HERNIA.

Section XXI.—Causes and Pathology.

	PAGE
Infantile cases—closure of the navel in development	238
Circumstances interfering with closure	240
Causes of the adult variety	241
Differences between the adult and infantile forms	242
Position of recti muscles, shape of hernial openings	243

Section XXII.—Diagnosis of Umbilical Hernia.

Symptoms and appearance in the child	244
,, ,, in the adult	246

Section XXIII.—Treatment.

Operation for radical cure by Dessault	247
Objections made by Scarpa	249
Advantages of a subcutaneous wire method	250

Section XXIV.—The Author's Operation for the Radical Cure of Umbilical Hernia.

Instruments used, choice of wire, &c.	251
Description of the steps of the operation	253
Dressing and after-treatment	256
Conditions favourable for operation	257

Section XXV.—Treatment of Umbilical Hernia by Pressure.

Objections to the ordinary apparatus	258
Description of the author's ring compress	259
Its action upon the hernial opening	260
Its use in irreducible cases	262

Appendix of Sixty Cases of Operation for the Radical Cure of Inguinal Hernia 263

ON RUPTURE.

INTRODUCTION.

LONG before the present time, the radical cure of hernial protrusions has occupied the attention of surgical minds of the highest order. Both in ancient and mediæval times, operations were done in great variety with a view to remedy what was then, as now, a very common failing of the powers of Nature in fulfilling the destiny of man to labour. The fatality attending some, and the inutility and inconvenient results of more of these operations, together with the improvement and extensive employment of a most useful palliative—the truss—have brought about a very general conviction, shared in by the most trustworthy authorities, of the folly and uselessness, and, by some, of the recklessness even, of any attempt to cure a disease which had been attempted so often in vain.

Rupture is not a disease directly or necessarily fatal, and, by care and the use of a truss, may be reduced to the category of a simple inconvenience. It is not, therefore, justifiable to induce a patient to undergo an operation which may cause pain, and even endanger life, in the uncertain hope of a permanent cure.

Such is the position held by some eminent in the profession

to be the judgment of medical ethics. Some would not even allow of such an operation being classed among the "operations de complaisance," upon which the surgeon may permit his patient, after having the case fairly stated, to judge for himself, but would heartily dissuade his patient from the step.

Like many other diseases which the profession has pronounced incurable, it has been the fate of ruptures to fall into the hands of unscrupulous men, who in every age have obtained a spurious reputation for curing a disease which is not to be distinguished except by experienced surgeons from many others easily curable, and concerning which it is not in some cases easy to decide whether it be cured or no. Both one way and the other the public have been repeatedly deceived, and professional men are naturally placed in an attitude of suspicion towards all such attempts to remove the stigma of impotence which in too many instances is fixed upon their art.

In these latter days of attention to the social sciences, the physical qualities of races and their deterioration are justly considered to be of paramount importance. Habits of active exertion and muscular training are encouraged with a view to military, scientific, mercantile, and colonizing capabilities. Doubtless, the frequency of ruptures, frequent as they already are, will be thereby increased until the habit has invigorated more than one generation, and impressed its influence upon their descendants. Thus has arisen an increasing demand for something to be done for ruptures leading to the revival of attempts, most of them futile, to cure this prevailing deformity. In America especially, where the frequency of rupture among the slave labourers has given a considerable money value to its cure, resuscitations of old operations, and a multiplicity of new methods have been practised, with, as far as I can learn, a want of success almost invariable. Many of them having a pretty well marked stamp of recklessness, have added to the distaste with which thinking men in this country view the revival of attempts towards the radical cure of rupture by operation.

On reading over the opinions of modern writers on this subject, one cannot but be struck by the importance they attach to the supposed dangers of meddling with the peritoneum and its offsets. Around this theory are grouped most of the objections made to operative interference. The theory alluded to seems to have been deduced from experience of operations performed upon this membrane in a state of disease or inflammation, as in strangulated hernia, or of operations exposing it extensively to external influences. Hundreds of operations implicating the healthy peritoneum, both upon herniæ and under other circumstances, without bad results, have been overlooked or ignored. This prejudice is, I believe, at the bottom of most of the objections; as it prevailed formerly against early operation in cases of strangulated hernia. In the latter cases it seems to have generally given way, rendering it more easy to be dealt with in the former class.

In a general way, inflammation of a parietal portion of the peritoneum has been confounded with that of the visceral layer, or general inflammation of the cavity near the important nervous centres. A secluded portion has been invested with the attributes of the whole; a logical error not uncommon. A closer examination of the fatal cases of attempted radical cure will show that they have mostly succumbed, not to primary peritonitis, but to the results of burrowing of matter, pyæmia, or wounded bowel. The danger of peritonitis cannot, therefore, be considered sufficient to deter the surgeon from endeavouring to perfect our means of curing rupture.

Having thus briefly alluded to the screens which obscure the view and cloud the judgment of the claims of hernia to a radical cure, we may endeavour to ascertain by the experience already obtained in what class of cases rupture should be placed by the surgeon, considering what is best to be done for his patient, and for our common humanity.

Rupture, especially inguinal, is a very common disease, more particularly in those classes of men who are the backbone and support of nations—viz., the labouring, military, and naval classes, the sons of adventure, exertion, and toil. The disease renders them incapable, for the most part totally, of the effective performance of their duties, and may place their lives in jeopardy under circumstances in which their greatest efforts are required, and surgical help, generally, not available. This result has been found by experience not to be effectually guarded against by the use of trusses, which often fail just when most needed, or cause so much trouble and expense that they are discontinued as unprofitable. Military and poor-law records show how frequently this is the case.

Before we can obtain a proper light from statistics upon this subject, we must compare the total number of deaths from attempted radical cure with that from strangulation and other fatal consequences of rupture, and the proportions of each to the total number of hernia cases. This alone will show whether society at large is a gainer by the radical cure of rupture.

In an individual case, upon the merits of which alone must the conscientious surgeon decide, it is futile to say—but this patient may never have his hernia strangulated; since it is easily answered—but also, he may have it strangulated. His habits, occupation, and position in life only exert their influence in rendering its occurrence more or less probable, not impossible. In a moment of extraordinary exertion, this result may occur to any hernia patient, and he may die from it.

Again, rupture is a very troublesome and incapacitating disease, rendering the patient's life uncomfortable, comparatively useless for bodily exertion, and not unfrequently miserable, exposing himself, and often a family, if in the lower ranks of life, to the perils of destitution. It is from this point of view that we may most justly consider the value of an operation for a radical cure to the patient himself, as regards its chances of success or failure in effecting its object. It is readily conceded that it is most difficult to obtain reliable statements of the ultimate results of these operations. Some which have been ushered into notice with most favourable accounts of success have been found to be attended with almost universal failure in other hands. Upon this point, therefore, it is incumbent upon the surgeon to speak from personal experience only, founded, as far as may be, upon cases openly treated in a public institution.

In the Appendix to the present work are inserted such cases only as have been operated on by the author himself, nearly all having had the guarantee of public observation. The total result is, that out of the sixty cases there given, treated by various modifications of the same principle, there have been eleven failures, two doubtful, four partially successful, and one death from pyæmia.

If these results are compared with the statistics of operations which are now considered by most surgeons as good and safe operations, they will not be found to contrast unfavourably.

Let us take varicose veins, hæmorrhoids, stricture, prolapse of the uterus and rectum, as affording a class of diseases admitting, like hernia, of both a palliative and radical system of treatment. Though but few of these are so severe a disability as a rupture, or so often dangerous to the life of the patient, yet they are now almost universally considered as fit subjects for operation with a view to a radical cure. Yet fatal casualties are not uncommon, while failures are certainly as frequent.

It is a question generally asked of the surgeon by his

patient, "Is the operation a safe and certain one?" The answer will vary somewhat with the experience, and more, probably, with the individual character of the surgeon.

The most reasonable definition of a safe operation seems to the author to be " an operation which has no peculiar dangers arising from the situation of the part, or the method of procedure, and which is not subject more than others to those accidental diseases which may occasionally follow any interference whatever with the surface of the body, such as erysipelas, tetanus, and pyæmia. The results above given will entitle the operations practised by the author to admission into this category.

Next,—is the proportion of success to failure such as to offer to the patient chances of cure which will overbalance the dangers and inconveniences of a rupture treated by a truss more or less efficient?

In estimating this, it must be borne in mind that in almost all of the cases called unsuccessful the patient is in a better condition than before the operation, inasmuch as a truss was rendered effective which had previously failed in keeping up the rupture. In none of the cases has the condition of the patient been rendered worse.

Many of the failures have resulted from causes which have led to various modifications of the proceeding. Some of these are better adapted for peculiar forms of the disease, others for the more simple varieties. That which I have found most generally applicable is the wire operation without compress, as given in the text. The great majority of the cases have been followed by such results as to afford great relief and satisfaction to the patient, and to raise the admiration of many experienced surgeons who have examined the patients from time to time.

By the methods followed by the author, and described in the following pages, we have the large percentage of forty-two out of sixty successful cases, *i.e.*, cases in which the use of a truss may usually, after a time, be dispensed with.

Of other methods which have latterly been practised, both in this country and abroad, the author has had experience only of the difficulty of obtaining accurate information as to their ultimate results.

In the hands of all surgeons from whom he could obtain such information, the general results have been very discouraging; and he has himself not been able to meet with a single case in which the patient has been able to go without truss with absence of protrusion one year after the operation, while he has had occasion several times to operate after other methods have failed.

It may be said that in an operation of this kind a certain tangible percentage of successful cases is absolutely necessary to induce patients to submit to the necessary inconvenience. The occasional occurrence of a fatal case, though somewhat discouraging, will, if the cause of death be not inherent in the nature of the operation, hardly constitute so formidable an objection as the chances of success being but slender. The numerous deaths from the use of chloroform but little affects its extensive employment, although its perils are well known to the public. The objection of a general want of success weighs very heavily against nearly all these modern operations.

The results of the cases appended to the following pages give both patients and surgeons ample encouragement to an extended trial of these methods of operating, and appear to the author to justify the inference that his operation for inguinal hernia is at the least as successful and as safe as any operation for the cure of prolapsed viscera, or that for varicose veins. Applied as it is to a deformity, more common, more dangerous to life, and more disabling, and affecting a class of persons to whom bodily strength and activity is of more vital importance to themselves and the country at large, it must be acknowledged that with an equal risk we have in this operation a greater power for good.

Again, the question arises:—In what kind of hernia, and in what class of persons, is an operation most desirable?

In old people, whose declining vital powers render them less capable of heavy labour or active exertion, and whose pursuits are generally of a light kind, and in whom any operation becomes proportionably hazardous, a radical cure of hernia is less imperative. Such protection as a good truss can afford is, except under peculiar circumstances, amply sufficient. After sixty we may consider such an operation generally unadvisable.

Females have, with some exceptions, lighter occupations than men, and may be sufficiently guarded against strangulation by a proper truss. They are, moreover, subject to the crural variety; which though happily less common, I consider, for reasons hereinafter given, less fitted for operative interference than inguinal. Except in market-women, milk-women, and others carrying loads habitually, and of a proper age, or in cases of inguinal hernia affecting the female sex, an operation is unadvisable. Inguinal hernia in females, however, may be more easily dealt with even than in males, on account of the absence of the spermatic cord and the complications which it involves.

In inguinal hernia affecting the young and able-bodied, who may at any time be called upon for such exertion as may endanger strangulation, a means of radical cure is a boon not lightly to be discarded; and as this is far more frequent than any other variety, indeed than all others put together, this desideratum is proportionately valuable.

Congenital hernia and the hernia of children are by many considered as perfectly curable by the pressure of a truss; but how common is it to find, even in these days of perfected trusses, a congenital hernia in an adult. The difficulties of adjustment, of constant and effective application, from the frequency of excoriation and abscesses, and of the supply of worn-out trusses among the poor, are almost insuperable, and in most cases prevent a cure. A simple, easy, and almost painless method of closing the inguinal canal in a child, totally free from danger, has surely as great a recommendation to the surgeon as the operations for hare-lip and club-foot, and is much more frequently required. Such an operation will be found described in the following pages, with cases in the Appendix. The results are so encouraging as to give a certainty and celerity to the process of cure, with which the tedious process by the pressure of a truss alone can by no means be compared.

To inguinal hernia and its varieties, then, is due the chief consideration with a view to a radical cure.

The presence of the spermatic cord, and the propriety of preserving intact the functions of the testicle, present the chief difficulties which beset the surgeon in effecting this end satisfactorily. The dangers will be easily avoided by surgical tact and anatomical skill.

The less frequency of crural and umbilical hernia renders them of less importance in a social point of view.

To the former, therefore, the chief part of the following treatise is devoted. Of the rarer forms of rupture, the experience of the author has not justified any especial mention.

PART I.

INGUINAL HERNIA.

I.

THE ANATOMY OF THE PARTS CONCERNED IN INGUINAL HERNIA.

It will be useful in this place to call attention to such particulars in the anatomy of inguinal hernia as are concerned in the operations about to be described, and necessary to be kept in view in considering the causes and treatment of the displacement.

An inspection of the surface of the groin before the removal of the skin will bring to notice externally a hard, bony projection. This is the anterior termination of the upper border or crest of the *ilium* or *haunch-bone*, and is called the *anterior superior iliac spine*. From this point downwards and inwards is a hollow groove curving gently with the convexity downwards, indicating the union between the trunk and leg at Poupart's ligament, and terminating internally near the genital organs at a slight eminence, composed chiefly of fat and covered with hair in the adult, the *mons veneris*. To the lower side of this may be felt another hard bony ridge, formed by the upper margin or *crest of the pubis or share-bone*. This presents, to a careful touch, a slight bony tubercle externally, projecting forwards, and called the *spine of the pubis*. Below the groove, upon the front of the thigh, is a triangular hollow forming the lower region of the groin, and hereafter to be described in connexion with crural rupture. Above is the rise or swell of the abdominal wall, projecting forward in a gentle curve. Close above the pubis, and a little external to the mons veneris, may be felt an opening in the deep structures. This is the *superficial or external abdominal ring*, through which, in the male, emerges a soft cylindrical cord-like mass, passing downward into the purse or scrotum to the testicle. This is the *spermatic cord*, containing the duct, vessels, and nerves of the testicle, which pass this way into the interior of the belly.

Under the skin of the groin the superficial fascia is arranged more distinctly than usual in two layers, between which are placed the cutaneous vessels and lymphatics. The *superficial* layer contains a loosely connected fat, and is continuous with the corresponding fascia on the abdomen and thighs. Towards the median line it is more plentiful and firm, and is composed of a greater proportion of fibrous tissue, contributing to make up the cushion of the mons veneris (Figs. 1 and 6, *a*).

The *deep* layer contains little or no fat, and consists of a very tough and elastic areolar tissue, with large and moveable spaces of a shiny bursal or synovial appearance. It is closely attached to Poupart's ligament to the front of the pubis, and to the iliac crest. Upon its elastic extensibility chiefly depends the facile gliding motion of the skin over the subjacent aponeurosis. In a normal condition of the parts a puncture or spot in the skin can be moved more than an inch in every direction (Figs. 1 and 6, *b*).

Both layers of fascia may be observed to pass down into the scrotum, surrounding the spermatic cord with a distinct tubular investment, within which it can be felt to move freely. This fascial tube is connected on the one hand with the mons veneris and investments of the penis, and on the other with the pubic spine and descending ramus. These connexions contribute to prevent any invagination of the posterior portion of the fascial tube into the inguinal canal beyond a certain extent; their recognition is consequently of some importance to the present subject.

Between the layers of fascia are placed the upper chain of lymphatic glands, which transmit the ducts from the penis and scrotum. These are arranged parallel with, and usually directly superficial to Poupart's ligament. (Fig. 1, *c*). Supplying them with blood, and crossing the groin obliquely upwards and inwards over Poupart's ligament, are three small branches of the common femoral artery.

The outer of these, the *superficial circumflex iliac*, is directed towards the superior iliac spine, about which it is distributed to the integuments (*d*). The central one—the *superficial epigastric*, crosses the inguinal canal a little external to its middle, supplying chiefly the glands, and is distributed to the integuments below the umbilicus (*e*). The inner one, the *superficial external pudic*, passes inwards and a little upwards to supply the mons veneris, and crosses the spermatic cord superficially as it emerges from the external abdominal ring (*f*). A fourth branch of the common femoral, the *deep external pudic*, perforates the fascia lata near the pubis, and crosses behind the spermatic cord, a little below the pubic spine, to reach the back of the scrotum and root of the penis, where it is distributed. In the operations for the cure of inguinal hernia, the only branches likely to be involved are the external pudic. They are usually so small as to require no ligature if divided, and give no trouble. Occasionally, however, one or other may be so enlarged as to give a considerable supply to the skin on the dorsum of the penis. It may then be felt pulsating over the hernial tumour, and require a ligature if divided.

At the root of the scrotum the subcutaneous fat suddenly ceases, and about an inch below the pubic spine is replaced by the unstriped muscular fibres composing the dartos, which completely lines the scrotal bag, mingled with a very tough, elastic, and vascular areolar tissue. A vertical septum of the same tissue forms a partition in the median line separating the testes. In the scrotum the layers of fascia are readily separated from the skin, the thin handle of a small scalpel being sufficient to detach almost any extent of fascia.

Fig. 1.

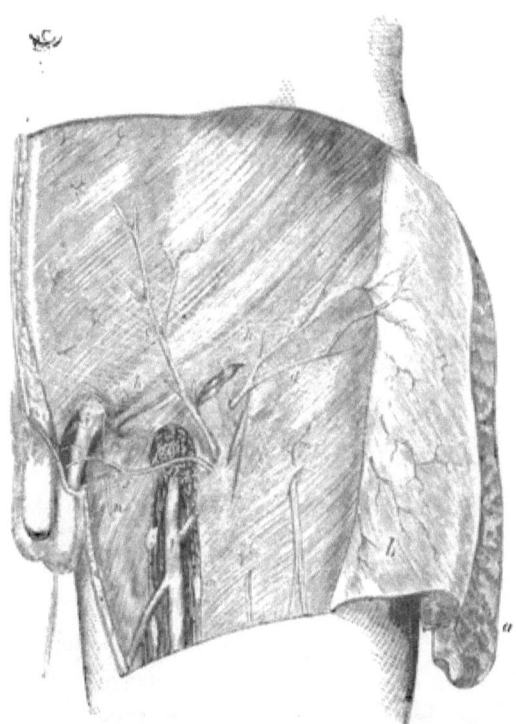

Superficial dissection of the inguinal and crural regions.

Below the level of the iliac crests the abdominal wall is made up of layers of muscular and aponeurotic structures. Most superficial is the strong tendon of the external oblique, the fibres of which curve downwards and inwards towards the pubis and median line, where they contribute with the other tendons to form a vertical line of union with those of the opposite side—the *linea alba*. Externally, towards the thigh, the fibres become thicker and more oblique, blending with the fascia lata, and with the deeper fascia to form the *crural arch* or *Poupart's ligament* (g). This strong band of fibres is arranged archwise between the anterior superior iliac spine and spine of the pubis, to which two bony projections it is firmly attached. It has a slight convexity tending downwards, outwards, and backwards, so as to form the hollow of the groin. Extension and abduction of the leg renders it tense by traction on the fascia lata, and through it compression is made upon the inguinal and crural rings by this position of the leg.

The fibres of the aponeurosis are bound firmly together by a thin, closely adhering layer of tough areolar tissue, which may be traced upwards to the fascia immediately investing the muscular fibres, and downwards into a fascia presently to be described as the *intercolumnar* (h). Openings of various sizes give passage to vessels and nerves lying in the abdominal wall. A larger and more important opening of a similar kind forms the *external abdominal ring* (i).

Immediately behind the aponeurosis of the *external oblique* are placed the *internal oblique* (Fig. 2, b) and *transversalis* (c) muscles with their conjoined tendon. Their muscular fibres lie externally, both taking origin from Poupart's ligament (a), the former extending down further and covering the latter, one occupying the outer two-thirds, and the other the outer third of the whole extent of the crural arch. Internally is placed their united or *conjoined tendon* (d), passing in front of the rectus and pyramidalis muscles to the linea alba and pubis. The *rectus* muscle (e) spreads out over the hypogastric region, and offers an effectual bar to hernial protrusion at this place. Towards the median line it closes up with its fellow muscle, separated only by the linea alba. Close to the pubis, the small *pyramidalis* muscle arising from the crest of that bone, and inserted into the linea alba for about two-and-a-half inches, renders tense the latter structure, and assists in closing the abdominal wall in front of the rectus tendon.

Posterior to all these, and lining the transversalis muscle and conjoined tendon, to which it is very closely adherent, is the *transversalis fascia* (Figs. 2 and 3 *f*). Internally, this fascia passes behind the rectus to the linea alba, forming with the subperitoneal tissue the only posterior covering of the muscle at its lower fourth, and continuous above with the posterior layer of the split tendon of the internal oblique which completes the sheath at the middle and upper parts of the abdomen. Externally, the fascia is continued downward below the edge of its muscle to be firmly attached to the deep surface of Poupart's ligament and to the fascia iliaca, and to form the internal abdominal ring and the anterior part of the crural sheath. At this part it is much thicker than above, and is further strengthened by scattered fibres derived from the tendon of the transversalis, forming an important barrier against direct hernial protrusion, and protecting the epigastric vessels which lie behind it, with the peritoneum and obliterated hypogastric cord.

The *inguinal canal* is an oblique passage through the layers of the abdominal muscles and fascia, varying from an inch and a half in the male, to two inches or more in the female. It is placed immediately above the inner half of Poupart's ligament. The superior or deep opening is situated about half or three-quarters of an inch above the centre of Poupart's ligament; and the inferior or superficial opening, immediately above the pubic spine into which that ligament is attached internally. These openings are called respectively the *internal and external abdominal rings*.

The *internal or deep ring* is an oval opening in the fascia transversalis, with its long diameter directed upwards and a little inwards (Fig. 3, *a*). Its inner and lower margins are tolerably sharply defined when viewed from the inner aspect. The outer and upper borders are less prominent, from the obliquity of the surface. From its margins is derived a thin funnel-shaped tube of areolar membrane, the *fascia propria vel infundibularis*, the deepest covering of the spermatic cord.

The *external or superficial ring* is an oblique triangular slit or opening between the fibres of the external oblique aponeurosis. The base of the triangle is placed above the outer half of the pubic crest, the superior angle is directed upwards and outwards (Fig. 1, *i*). A line bisecting the triangle would be placed rather more vertical than Poupart's ligament. The sides of the slit are arranged in two bands called *pillars* of the ring, the inner of which is also superior, and the outer, inferior, from the obliquity of the opening. Crossing from one to the other pillar in front of the ring, and continuous with the fascia before mentioned as closely investing the aponeurosis, is a tolerably strong layer of closely woven tissue called the *intercolumnar fascia* (*h*). At the upper part of the ring are developed in it strong bands of fibres, which, connected externally with the outer half of Poupart's ligament, arch transversely over the opening in curves having the convexity downwards and outwards, and which, separating from each other as they pass inwards, become finally directed upwards, and are lost in the aponeurosis near the linea alba (*k*). These bands have received the name of the *arciform fibres*. At the lower part of the ring the intercolumnar fascia is prolonged downwards over the cord into a tubular investment which has been named the *fascia spermatica externa* (seen at *i*). To the yielding of the intercolumnar fibres it is that dilatation of the external ring by a hernial protrusion is chiefly owing. When deprived by close dissecting of this investment, the pillars of the ring are fairly exposed. The *internal pillar* is flat and riband-shaped, and composed of parallel fibres. A little way from the border of the ring, at an oblique line passing downward and inward to the pubis, it becomes blended with the subjacent conjoined tendon of the internal oblique and transversalis muscles. It gives off many fibres into the tissue forming the mons veneris (some of which join the suspensory ligament of the penis), and is implanted below upon the front of the angle of the pubis and anterior ligament of the symphysis, at a little distance below the crest of that bone.

The *external pillar*, triangular in shape, with the apex downward and implanted upon the pubic spine, is connected externally with Poupart's ligament, with the insertion of which it is so blended as to be identical, forming a rounded tendon very evident to the touch. To form this tendon the fibres of the pillar converge to those of Poupart's ligament. The strength and width of this pillar vary much in different subjects. In some it is thin and weak, and lost in Poupart's ligament at some distance from its attachment. Near its lower end, it is grooved above by the passage of the spermatic cord, which crosses it obliquely just above and external to the pubic spine. The presence of a hernia tends to evert the border of the pillar, and to diminish its width.

Placed immediately behind the aponeurosis forming the outer pillar at the upper third of the inguinal canal, are the lower muscular fibres of the internal oblique, arching over the cord obliquely and covering the internal abdominal ring (Fig. 2, *b*). In well-developed subjects, the lower part of this muscle is thicker and more bulky than the upper, and often separated from the rest by a distinct cellular interval, below which it is strengthened to resist protrusion through the internal ring.

Passing downwards from behind it, and blended more or less both with it and the transversalis muscles, are the muscular fibres of the *cremaster* (Fig. 2, *h*). These are attached externally to the lower third or more of Poupart's ligament, pass downward through the external ring, and form distinct bundles or loops over the cord in the scrotum. The bulk of the muscular fibres lie on the outside of the cord upon Poupart's ligament; the lower become lost on the tunica vaginalis; the

Fig. 2.

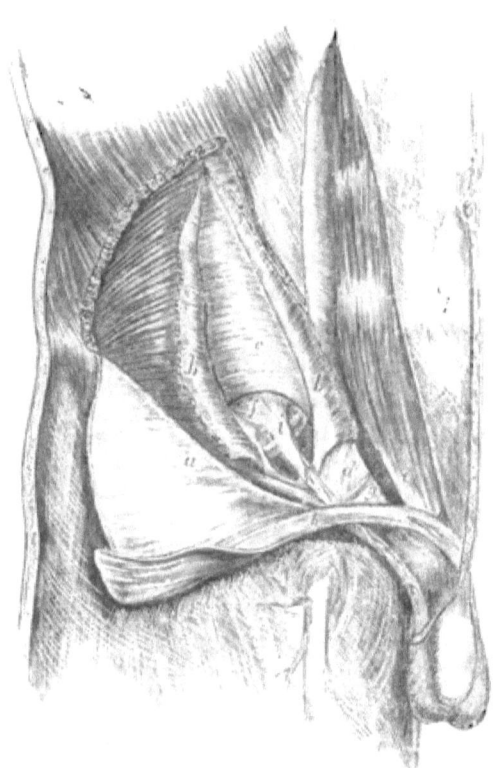

Deep dissection of the inguinal canal and abdominal wall.

upper re-enter the external ring, and become implanted upon the conjoined tendon; while the middle set of loops become lost in the fascial tube which connects the different bundles together into a continuous layer called the *cremasteric fascia*. This investment of the spermatic cord lies between the intercolumnar fascia and fascia propria, and forms the chief thickness of the deeper coverings of the cord.

Lying upon the fibres of the internal oblique and cremaster muscles is seen the termination of a small nerve, the *ilio-inguinal*, a branch of the first nerve of the lumbar plexus. This nerve, after perforating the deeper muscular fibres near the iliac spine, pierces the intercolumnar fascia at the lower part of the external ring (at *i*, Fig. 1), and is distributed upon the upper and inner part of the thigh and adjacent scrotum. The accidental inclosure of this nerve in the ligature for the cure of hernia occasionally causes pain along the iliac crest, which may give rise to unfounded apprehensions of mischief.

The structures just described make up the *anterior wall* or boundary of the inguinal canal,—intervening between it and the integuments.

The *posterior wall* or boundary is formed by three layers of fibrous and areolar tissue intervening between the spermatic cord and the peritoneal structures.

The deepest and most extensive of these is the before-mentioned *fascia transversalis* (Figs. 2 and 3 *f*). Thin and unimportant at the upper part of the abdominal wall, where it lines the transversalis muscle, towards the groin it becomes much thicker and denser by the addition of fibres from the tendon of the transversalis, and is connected closely with the deep surface of Poupart's ligament, with the iliac fascia, and with the outer border and deep surface of the conjoined tendon. Here it forms the oval opening of the *internal abdominal ring*, and gives off over the cord, the funnel-shaped investment called the *fascia propria* or *infundibular fascia* (Fig. 2, *i*). This fascia covers and obscures in front the edges of the ring from which it springs, and the spermatic cord, which it invests as it passes through the ring. The best view of the ring and the derivation of the fascia is consequently to be obtained from within, where the fascia will be seen to present a prolongation similar to that of the finger of a glove from the hand cavity (See Fig 3, *l*).

The connexion of the fascia transversalis with Poupart's ligament forms a groove for the lodgment of the spermatic cord. This groove is deepest above, opposite to the internal ring, at which point the finger of the operator, invaginating the scrotal fascia, can be placed easily between the cord above, at the internal ring, and Poupart's ligament below. The groove becomes shallower downward, and terminates upon the external pillar itself near the pubic spine.

The union of this fascia transversalis with the iliac fascia around the iliac vessels, forms the crural sheath to be hereafter described in relation to crural hernia. It covers and protects the external iliac vessels and their epigastric and circumflex iliac branches, all of which lie behind it.

In the operations for inguinal hernia hereafter described, it is of great consequence to recognise the fascia transversalis in the above important relations and functions, as affording a guide to the finger and a protection to the vessels and peritoneum.

Placed in front of, and blended with, the last fascia at the inner two-thirds of the posterior wall, is the *conjoined tendon* of the internal oblique and transversalis muscles (Fig. 2, *d*). At its upper border the tendon is connected with the muscular fibres of the transversalis, which terminate in it about three inches above the pubis. Upon its anterior surface are implanted the fibres of the internal oblique, which, continued lower down than those of the former muscle, are attached in bundles, at varying distances from the linea alba, forming an oblique line of attachment downwards and inwards. Below these are inserted, by a cellular attachment, a few of the fibres of the cremaster muscle at their inner extremities, after passing in loops over the cord and through the superficial ring (*h*). From the line of attachment of the internal oblique muscular fibres inwards, the tendon is much thicker and stronger. Externally, the tendon is thin and its fibres are scattered, becoming apparently lost in the fascia transversalis. Internally, it passes in front of the rectus and pyramidalis to the linea alba, blending with the aponeurosis of the internal pillar of the outer ring, and with the triangular tendon. Below, it is connected firmly with Gimbernat's and Poupart's ligaments, with which it is implanted into the pectineal line and spine of the pubis.

Directly opposite to the external abdominal ring, this tendon is strengthened and much thickened by the superposition of a triangular layer of fibrous tissue passing across the linea alba from the external oblique aponeurosis of the opposite side.

These fibres, after decussating with those of the internal pillar, pass outwards and downwards to the pubic symphysis, crest and spine, and sometimes as far as the pectineal line, where they are implanted with those of the conjoined tendon. This structure is best named the *triangular aponeurosis* (Fig. 2, *g*). It is more conspicuous in some subjects than in others, and usually more in females than in males, especially those who have borne children. In some it is so much blended with the conjoined tendon as to be almost indistinguishable, except by a perceptible thickening. It contributes an important aid in resisting a direct protrusion through the external abdominal ring at the outer edge of the rectus.

The conjoined tendon and its associated structures may be raised from the deeper tissues, and its obscured external border rendered salient and perceptible to the touch, by passing the finger along the inguinal canal, and raising forwards the lower fibres of the internal oblique muscle upon it. The edge of the tendon will then be felt on the inner side of the finger.

The *superior boundary* of the canal is formed by the apposition of the several layers of muscles and fasciæ, between which the canal is placed. Crossing obliquely from its attachment to Poupart's ligament in front of the canal to its insertion upon the conjoined tendon, and arching over the middle of the canal in the manner of a skew-bridge, is the lower border of the internal oblique muscle (*b*). This peculiar obliquity is such, that the finger of the surgeon inserted along the inguinal canal when dilated by a hernia, passes behind the muscle at the upper part, but in front of its tendon at the lower part of the inguinal tract. Its position would thus limit the upward dilatation or rise of a hernial tumour, which it would at the same time closely embrace on the upper and anterior aspects.

The *inferior boundary* of the canal is formed chiefly by Poupart's ligament, along its inner half, and partly by the attachment of the fascia transversalis, conjoined tendon, and triangular aponeurosis to its deep surface, forming together a deep groove which lodges the spermatic cord, and which terminates below upon the external pillar of the superficial ring.

Bounding the inguinal canal at its upper and outer extremity, and indicating the position of the internal ring, not otherwise readily distinguished from without, is the lower border of the *transversalis muscle* (*c*), passing horizontally to be connected with the conjoined tendon. In some cases this muscle does not descend so low as the ring, a condition usually accompanied by a similar want of development of the internal oblique, and indicating a hernial predisposition.

The inferior termination of the canal is indicated on its deeper aspect by the edge of the tendon of the *rectus abdominis muscle*, lying behind the conjoined tendon and internal pillar of the outer ring. A close inspection of this muscle will show that its origin from the pubis is arranged into two tendinous portions connected by areolar tissue. The inner one is more bulky, and more or less intermixed with muscular fibres, and arises from the superior ligament and fibro-cartilage of the pubic symphysis; the outer one is thin, entirely tendinous, connected by its thin outer border with the fascia transversalis, and implanted below into the pubic crest (see Figs. 2 and 3 *e*). Above the tendon, the belly of the rectus muscle in well formed subjects curves boldly outwards and upwards towards the false ribs. In hernial subjects it may be distinctly felt from the inguinal canal, and in many, the finger may be pushed completely behind it, while contracting, without rupture of tissue.

If the peritoneal cavity be opened by a transverse incision from one iliac spine to the other, joined by a vertical one down to the pubis symphysis, on the bowels being drawn aside, the serous pouch in the groin will be seen to present a dimple opposite to the internal ring, where, upon close inspection, faint traces of a cicatrix may be observed, indicating the closure of the *canal of Nuck* or vaginal prolongation of the peritoneum, after the descent of the testis. Below it is a smooth line, indicating the position of Poupart's ligament and separating it from the depression over the crural ring still further down. To the inner side, the serous membrane is slightly raised in a line directed upwards and inwards, by the epigastric vessels. On the other side of this line another slight depression is seen, which is bounded internally by a still more prominent ridge, also directed upwards, inwards, and forwards. This latter ridge is caused by the obliterated cord of the hypogastric artery, the umbilical branch of the internal iliac in the foetus (Fig. 3 *i*). This, after curving forward along the side of the bladder, ascends above the fundus of that viscus, and contributes to form the border of its *superior false ligament*. The chief portion of this latter compound structure is made up of a very loose fold of peritoneum of a triangular shape, having its base at the bladder and its apex at the umbilicus, and containing in the median line the remains of the obliterated urachus, and a quantity of very loose areolar tissue and fat. Its office is to form an investment to the fundus of the bladder, when so much distended as to rise out of the pelvis.

On raising the peritoneum, its superficial surface is seen to be connected to the fascia transversalis in the groin by a very lax and plentiful areolar tissue containing much yellowish fat, which, on being seen against the greyish blue of the serous membrane and viscera, obtains usually a tinge of greenish grey. This has received the name of the *subperitoneal fascia*. It is especially plentiful in the superior ligament of the bladder and behind the conjoined tendon, less so on the outer parts of the groin. In the former situation it permits the finger to be passed easily between the conjoined tendon and the peritoneum. It is here joined by a deep layer from the fascia transversalis, with which it forms the posterior part of the sheath of the rectus at this point.

Lying in this fascia, behind the fascia transversalis, and internal to the internal ring, passes the *epigastric artery* on its way upwards and inwards to the sheath of the rectus.

The *external iliac artery* (*a*), of which it is a branch, is placed a little external to and about an inch below the internal abdominal ring, upon the border of the psoas muscle, along the side of the pelvic brim. The vein is to the inner side and a little below its level. They are covered by a sheath, divided by a septum, and derived from the subperitoneal fascia which binds them down to the fascia iliaca. Internal to them, and considerably below their level, are placed the glands which transmit the lymphatic ducts from the crural ring to the aortic chain (Fig. 3, *k*). Lying upon the artery is the genital branch of the genito-crural nerve on its way to join the spermatic cord at the internal abdominal ring; and crossing superficial to it, is the vein of the circumflex iliac branch passing inwards, and the vas deferens directed upwards and outwards, on its way from the pelvis to the internal abdominal ring (*g*).

Close above Poupart's ligament the artery gives off two considerable branches, the *deep circumflex iliac* outwards and upwards, and the *deep epigastric* upwards and inwards.

Fig. 3.

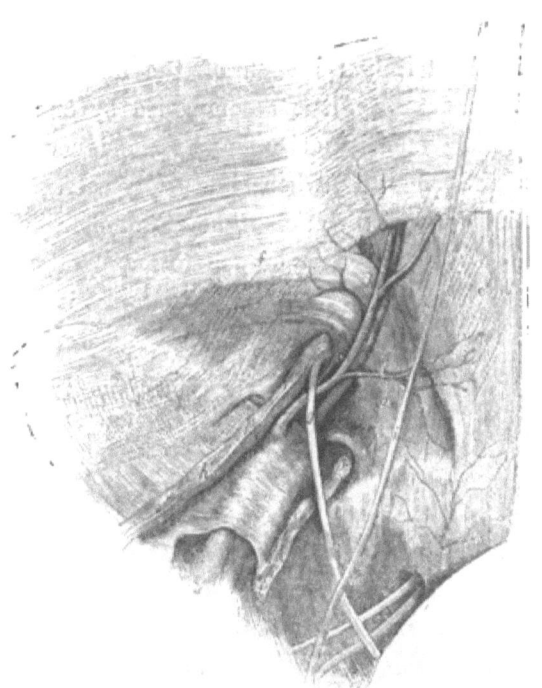

Dissection of the parts of inguinal and crural hernia from the deep or internal surface, the peritoneum and its fascia removed.

The former, rather the larger of the two, lies behind the iliac fascia along the deep surface of Poupart's ligament, passing towards the iliac crest, above which it anastomoses with the lumbar and gluteal arteries (Fig. 4, *d*). A considerable branch often passes upwards and inwards close to the outer side of the internal ring between the internal oblique and transversalis muscles, to anastomose freely with the epigastric and intercostals.

The *epigastric artery*, of the size of a small quill (Fig. 3, *b*), passes across, behind Poupart's ligament, between the internal abdominal and crural rings, to the lower and inner side of the former, whence it ascends obliquely upwards and inwards towards the sheath of the rectus, which it enters at the *fold of Douglas* (*c*), to be distributed to the muscle and to anastomose freely with the terminal branch of the internal mammary. Like the circumflex iliac, it is accompanied by one and sometimes two veins; one is usually found to its inner side. The branches of the epigastric, when abnormally enlarged, may become very important in relation to hernia of both kinds.

1st. *The pubic.*—This is a branch to supply the symphysis pubis, and to anastomose with the obturator as it passes under the subpubic notch. This branch is very often enlarged so as to substitute the typical obturator from the internal iliac. It then constitutes a very important and dangerous relation to crural hernia. 2nd. The *cremasteric,*—given off at the internal ring, perforates the fascia transversalis, and supplies the cremaster muscle, anastomosing freely with the spermatic. Occasionally I have found it enlarged, and supplying the place of the latter. *Lastly,*—there are *superficial,* perforating the aponeurosis with the cutaneous branches of the intercostal nerves, and—*internal* and *external* branches of anastomosis with the intercostals, the circumflex iliac, and the opposite and superficial arteries of the same name.

The course of the epigastric artery, upwards and inwards towards the border of the rectus muscle, will be seen to form the external boundary of a triangular space formed between it and the border of the rectus tendon, which has for its base below, the spine and pectineal line of the pubis and the attachments of Poupart's and Gimbernat's ligaments to these points of bone. This triangle (Fig. 3, *d*), which being particularly described by *Hesselbach*, has been called by his name, is filled up or closed in by the deeper wall of the inguinal canal, which is made up for a short space near the artery, externally, by the transversalis fascia only,—at the inner two-thirds, by the addition of the conjoined tendon,—and at the inner third, by the additional superposition of the fibres of the triangular aponeurosis, which strengthen it superficially opposite to the external abdominal ring. Passing upwards and inwards, across the deeper surface of this space, is seen the obliterated cord of the hypogastric artery (*i*). This, enveloped in the outer border of the superior false ligament of the bladder, is placed a little outside the vertical centre of the conjoined tendon. About the inner half of this tendon, then, is in apposition behind to the superior false ligament of the bladder, when that viscus is moderately distended.

The *vas deferens*, or spermatic duct (*g*), after crossing the external iliac vessels on their inner side, makes a sharp curve round the epigastric vessels to enter, through the internal ring, the inguinal canal, where it lies to the inner and back part of the cord.

On its outer side are the *spermatic plexus of veins* with the *artery* and *nerves* (*h*) which, after crossing the psoas and iliacus muscles, enter the canal, lying external and superficial to the vas deferens. At the inner aspect of the internal ring itself may be seen a portion of the subperitoneal fascia (*l*), holding together the constituents of the cord, uniting them slenderly to the depressed cicatrix of the peritoneum, and connected with the margins of the internal ring. These margins will be seen to be more prominent internally and below, near the epigastric vessels and vas deferens, and to give off anteriorly the thin areolar infundibuliform prolongation over the cord.

The *spermatic cord* effects its passage through the abdominal walls in an oblique direction from above downwards, and from without inwards. The *inguinal canal* thus formed is an oblique valvular passage, with its upper end placed deeply and externally at the internal ring, and its lower superficially and internally at the external ring. Consequent upon this oblique course its anterior wall is thicker above than below, and its posterior wall thicker below than above. Thus is formed a true valve which, in its normal condition, resists powerfully any irregular protrusion of the abdominal contents. At the upper part of the canal the cord is placed at the distance of half or three-quarters of an inch above Poupart's ligament; but at the lower end it lies directly upon that structure. Thus the course of the cord through the canal is oblique in a double sense, viz., it is oblique from above downward and inwards, as well as from the deeper to the more superficial parts. It is doubly oblique, also, in reference to the *axis of the canal itself*, as the canal is to the abdominal wall in which it lies.

During its passage, it has obtained, at short intervals, a thin covering from each of the layers of fascia and muscle which it has traversed, viz., 1st,—from the fascia transversalis at the borders of the internal ring; 2nd,—a thicker layer, containing the muscular loops of the cremaster, from the lower border of the internal oblique and transversalis muscles; 3rd,—from the intercolumnar fascia at the borders of the external abdominal ring; and lastly,—a thick investment from the superficial fascia.

Inclosed in these coverings, the constituents of the cord itself, composed of the spermatic duct, vessels, and nerves, are connected loosely together by cellular tissue formed of the remains of the *canal of Nuck*, which, in fœtal life, connects the peritoneum with the cavity of the tunica vaginalis, and in adult age can be distinguished in the areolar connexion of the cord with the digital impression and cicatrix on the peritoneum at the internal ring.

The spermatic cord is freely moveable in the canal, and its constituents may generally be distinguished, and sometimes considerably separated, by the finger, when pushed through the external ring. The vas deferens is known by its whipcord-like feel, lying internal and posterior to the vessels. In old hernia cases, the connexions of these different parts of the cord to the walls of the canal are usually looser, giving a greater mobility to the structure, and enabling the surgeon easily to turn them aside with the finger. When the spermatic veins are varicose they are still more easily distinguished. This condition, also, often is present in old ruptures. The anatomy of the parts concerned in inguinal rupture is more free from irregularities than in most other regions. The arrangement of the epigastric vessels is not commonly found subject to variation, and they may always be calculated upon in their relation to the deep inguinal ring, whatever be their origin from the parent trunk. The irregular origin of the obturator, described more particularly in the parts of crural hernia, does not affect the relations of the inguinal canal. A want of development of the lower part of the internal oblique and its conjoined tendon, sometimes present, has been before referred to.

II.

CAUSES AND PATHOLOGY OF INGUINAL HERNIA.

Two theories have been broached as to the cause of this complaint; and as the adoption of one or other may influence the judgment as to the curability of the disease, it is necessary briefly to notice them in this place. One, which has received the support of Warton, Morgagni, Brendel, Richter, and others, and which is now held by some surgeons, is, that the immediate cause, of inguinal hernia especially, is an abnormal elongation of the mesentery, permitting such an extent of play to the bowels as to allow of protrusion through the openings of the groin, under the pressure of the abdominal muscles. The assumption is, that a mesentery of proper length would not allow any protrusion of the intestines through an opening in the abdominal wall at this point.

Without recapitulating the able arguments in refutation of this opinion, from the eminent pens of Scarpa (*Traité pratique des Hernies*), and of Samuel Cooper (*Surgical Dict.*), and others, some facts may be mentioned which any anatomist may observe, and which tend to support the conclusion of the first-named surgeon: "that the necessary elongation of the mesentery does not precede the displacement of the intestine, but is more probably simultaneous with it, and caused by the dragging of the already displaced bowel."

When the bowels are distended with food or air, the whole front wall of the abdomen is projected forward. There being no vacuum, and action and reaction being equal, the pressure is equally distributed over the whole of the containing parietes. The distension of the bowel must stretch the mesentery to its utmost, one fold pushing another before it. If the sides be sound, and equally resisting, the whole of the flexible part of the abdominal walls yields equally to the pressure, and no hernia occurs; but if one part be weak while another is resisting, it yields to the pressure at the time of its greatest force. This culminating point is produced by the action of the recti, and other abdominal muscles, antagonizing the downward force of the diaphragm and the inspiratory action of the lungs. The effect of this combination in producing protrusion is seen in cases of wound of the abdominal wall passing into the peritoneum. Wherever may be the site of such a wound, protrusion of bowel invariably occurs, if the latter be in a state of distension.

The most likely, as well as the most frequent, place for the abdominal walls to yield before such pressure is at the sides of the recti muscles in the aponeurotic structures of the groin, especially when Nature has not sufficiently fortified the internal abdominal ring. But often the umbilical opening yields, especially in children, being at that time less firmly closed. This opening is placed nearer to the point of attachment of the mesentery than those of the groin. In some cases the obturator opening, the vagina or the sciatic notch, are found to be the weak part, although these are placed much further from the root of the mesentery than the groins.

The comparison of these facts leads to the induction that the vice is essentially in the containing walls, rather than in an elongated mesentery.

We find in the dissecting-room that there is a great variety in the position of the attachment of the mesentery to the spine. It is often attached to the posterior part of the abdominal wall one or two vertebræ lower than usual.

It is thus very common to find a great part of the small intestines lying in the cavity of the true pelvis, between the bladder and rectum. This result may also be brought about by a hypertrophy of the liver, stomach, or spleen. And yet hernia is by no means more frequently associated with these conditions, because the abdominal walls may be efficiently resisting at every point, and yield equably before the pressure. If the mesentery, then, be long enough to allow the small intestines to lie in the true pelvis, *a fortiori*, it is long enough to permit of protrusion at the groin, in ordinary cases.

Again, the direction of the mesentery is towards the left side of the abdominal cavity, and the small intestines lie chiefly in the left lumbar and iliac and hypogastric regions. On the left side, also, the mesentery is considerably longer than on the right, though its point of attachment is higher. If the supposition in question were true, hernia should be more common on the left than on the right side, whereas the precisely opposite is the fact.

In cases of hernia irreducible from strangulation or adhesion to the sac, post-mortem examination usually shows that the mesentery or omentum are dragged down with the bowel and pulled upon, by that adherent or strangulated part *only*, the rest of these membranes remaining of their normal length, and considerably shorter than the retained portion.

But perhaps the most conclusive argument of all is to be drawn from the fact that we find hernia to be most common, not in subjects with elongation of the mesentery, but in those with deficient or peculiar formation in those parts of the abdominal parietes in which the hernia most commonly occurs.

It will be at once seen by the reader, that if hernia be primarily and principally owing to abnormal elongation of the mesentery, any attempt to cure it by closing up the opening, and strengthening the abdominal wall, must be futile, or result in a protrusion elsewhere. It is such a notion which has led to the popular opinion that the cure of a hernia on one side will lead to its production on the other. This argument has been sometimes urged against the use of a truss even on one side only. The real fact, however—as those who have seen the most hernia subjects will the most readily allow—is, that double rupture is as commonly found where no truss at all has been used—such as congenital cases; or where, on the side first affected, the rupture has become irreducible, rendering the retaining power of the truss ineffective.

This much, however, may be granted to the theory under consideration, that a general laxity of the intestinal attachments to the posterior wall of the abdomen, coinciding with a similar condition of the abdominal walls themselves, will predispose in some measure to the production of a rupture.

But, fortunately for the patient and the reputation of surgery, the main cause of the evil lies more within the reach of remedy.

The most eminent authorities on this subject are unanimous in support of the second theory alluded to,—that the chief cause of hernial protrusion lies in a deficiency in some part of the walls containing the intestines. As to the precise structure in which the retaining power resides, there is again some difference of opinion.

Some suppose that the parietal peritoneal layer, spread over the inner surface of the groin, is the most powerful agent for retaining the viscera opposite the tendinous openings.

In a perfectly healthy subject, the peritoneum, as well as the other serous membranes, is more tough, elastic, and resisting than its thin, transparent appearance would at first sight indicate. At the internal abdominal ring there are evident traces of a cicatrix closing up the canal of Nuck. This, as before seen, is closely attached to the margins of the opening in the fascia transversalis, and continuous with the tissue which binds together the elements of the cord. It must evidently afford considerable resistance to a protrusion, aided as it is by the slippery smoothness of the serous surface upon which the intestinal pressure slides. A sense of something giving way, which is experienced on the first occurrence of many ruptures, is probably due to yielding of this resistance coinciding with a forcible dilatation of the internal ring. Sometimes such yielding may be due to a degeneration of the cicatrix: such comparatively adventitious tissues are usually the first to degenerate. When a pouch is once fairly formed, the process of dilatation begins; such a pouch, however, cannot be formed until the internal ring is more or less dilated. Until a rim be formed round the sac, the intestine cannot fix itself for dilatation, but slips off, and expends its force equally upon the surrounding parts. Digital impressions without distinct margin are common enough in the parietal peritoneum, unattended by hernial protrusion.

In the immediate neighbourhood of the internal ring, the serous membrane forms a loose fold, the superior false ligament of the bladder. This is attached to the muscular wall by tissue so loose and yielding as to afford space and play to the bladder for its varying volume. When the connexion of the peritoneum to the margins of the ring is detached, here is ample supply of serous membrane to form one sac or more of any ordinary dimensions.

For cases of direct hernia the supply of sac is still more ready, many of such cases being formed entirely within the limits of the bladder-fold of peritoneum.

Though weakness and general laxity of the peritoneal investment may, then, predispose to the formation of hernia, and, in common with a general laxity in the connexions of the abdominal layers, facilitate a rapid production and growth of hernial tumours, yet that this is the chief cause of hernia, as some suppose, can scarcely be maintained, seeing that hernia frequently occurs in situations where no such laxity is present. The lining membrane of the arterial system does not form the chief resistance against aneurismal protrusion, nor is the synovial membrane of the joints their principal bond of union.

That the sac of a hernia, however, when thickened by friction, pressure, and inflammatory deposit, may form a barrier to further distension, is shown by the dense ring which is formed at its neck in many old cases. In some it has been found to have resisted further dilatation so much that a fresh sac has been formed above it, and the unyielding ring displaced downwards bodily, so as to form an hour-glass constriction internally.

In a few cases this separation of the sac into distinct compartments has been found to be complete, affording a valuable hint, not only as to the inefficiency of a mere adhesion of the sides of the sac alone in preventing a further escape of sac and intestine, but also on the possibility of inducing, by external means, such action in the sac cavity. Sir Astley Cooper, in his valuable work on *Hernia*, figures two cases of this kind, in which the division of the sac was placed at the external abdominal ring, converting a scrotal hernia into a bubonocele (*Plate* V.) In one case serous effusion had occurred in the lower or closed sac. In Fig. 3 of the same Plate is given a representation of a hernial sac interrupted in its formation by many incomplete septa, each of which had evidently at one time formed the internal opening of the neck of the sac into the peritoneal cavity, and become in its turn displaced downwards by a fresh protrusion of sac above under increased pressure. Here, as often, Nature points out a means to help us in the more perfect construction of a barrier, which has failed only for want of further combination with the tendinous sides of the hernia track high up the canal.

The chief cause of rupture, then, consists in the inefficiency of the tendinous or muscular containing walls of the abdominal cavity in resisting the pressure from within.

The cause of inguinal hernia lies in the failure of the valvular action of the walls of the canal. According to the way in which this is effected, the hernia is called *oblique* or *direct*

Fig. 4.

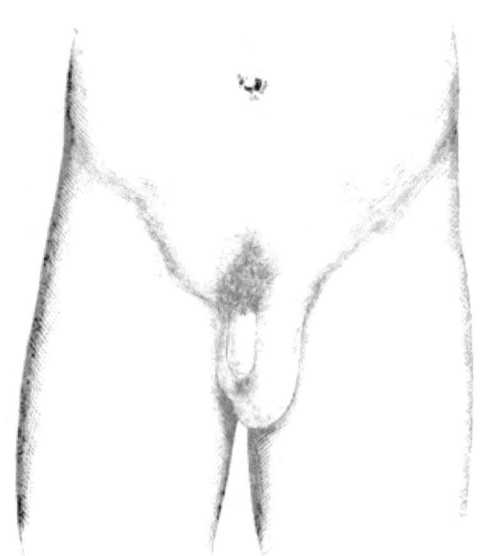

inguinal hernia—the one following throughout the oblique course of the inguinal canal, and the other passing more directly through the external abdominal ring. Both of these, in extreme cases, descend into the scrotum.

In the adjoining figure (Fig. 4) is represented, on the right side, the external appearance of an *oblique inguinal* rupture, while yet in the limits of the canal, in the condition called *bubonocele;* and on the left side, that of a **direct** hernia which has passed into the scrotum. In *oblique hernia* it results primarily from a simple dilatation of the deep ring, followed by that of the walls of the canal, the pillars of the superficial ring, and, in the male, of all the coverings of the spermatic cord which invest it, in regular succession.

In *direct hernia* it is formed by the giving or splitting of the fibres of the conjoined tendon and fascia transversalis, internal to the epigastric artery. The inguinal canal is entered from the side by a deficiency in the deeper of its valvular walls. In both, the dilatation of the canal and outer ring is a secondary, and often a protracted process. The weakness and tenacity of the layers, the laxity of their connexions, the degree of muscular power residing in the abdominal walls, the prevalence of a distended condition of the intestines from digestive causes, and the extent of deposit of adipose tissue in the peritoneal appendages, all contribute—by the varying amount of disturbance of the balance between the protruding and resisting forces—to facilitate or retard the formation of the sac.

The form assumed by the internal opening is always more or less circular, from the equable nature of the fluid pressure of the intestinal contents, or of the soft, loose fat of the omentum.

In *oblique* hernia, the deep epigastric vessels, even in an early degree of distension, are in close relation, internally and inferiorly, with the neck of the sac, which is, in fact, half surrounded by them. The larger the internal opening becomes, the more closely is the neck of the sac embraced by these important vessels, and the more near is the centre of the opening brought to the edge of the rectus muscle. The dilatation is accomplished chiefly at the expense of the inner and lower borders of the ring, until in large cases of scrotal hernia, the inner and outer openings are brought fairly opposite to each other, and the case assumes the appearance of a direct hernia. It has, however, this important difference—

that the neck of the sac is three parts surrounded by the epigastric vessels which are found below, internal to, and sometimes even partly enveloping, the upper margin of the opening. In the early stage, the protrusion through the internal ring is resisted by such of the lower fibres of the internal oblique as may be placed in front of the ring, and by the strong aponeurosis of

Fig. 5.

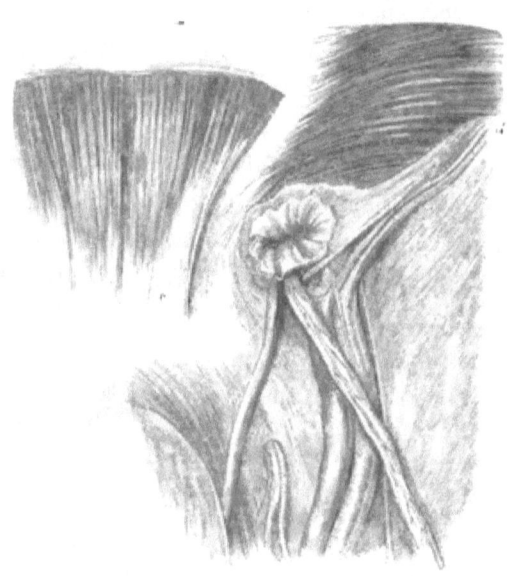

Dissection from the peritoneal surface of the parts affected by an oblique rupture. The peritoneum, its fascia, and the fascia transversalis, except a small portion around the neck of the sac, are removed; the sac is cut off at its neck in the deep ring; the epigastric artery is seen in its relation below the neck, at the inner side it is removed to show the deep aspect of the conjoined tendon (*h*).

the external oblique forming the upper part of the outer pillar of the superficial ring. The resistance of these structures turns the direction of the rupture downwards and inwards towards the outer ring; while the interval between the cord and Poupart's ligament at this point is dilated into a cavity lodging the sac, by the protrusion of the superficial wall forwards, and the deeper wall backwards. If the more superficial structures be very resisting, little tumefaction is produced at the surface, the yielding of the deeper wall forming the greater part of the space for the sac.

In *direct hernia* the protrusion occurs in the triangle of Hesselbach (see *d*, Fig. 3, page 25); the epigastric artery is consequently placed to the outer side and above the neck of the sac. Not unfrequently, the protrusion occurs on the inner side also of the obliterated hypogastric cord which bisects this triangle, as before described; the serous membrane forming the sac being derived entirely from the superior false ligament of the bladder. In such cases, the hernia is rapidly formed, and increases to a very large size, partly from the abundance of the supply and the laxity of the serous membrane in this place, and partly from the directly opposite position in which the internal opening is placed with regard to the superficial ring. In these ruptures the finger, passed into the neck of the sac, can feel the edge of the rectus muscle bounding directly the neck of the sac internally, while the conjoined tendon is not distinctly perceptible. The functions of the bladder are more interfered with, and in large cases the fundus of this viscus has been found involved in the hernia. More frequently, however, the protrusion of the rupture takes place between the epigastric artery and the hypogastric cord, in the depression which there normally exists on the surface of the peritoneum. In all cases, however, of direct hernia, the protrusion is somewhere within the limits of the triangle of Hesselbach, and placed inferior and internal to the internal abdominal ring,—entering the inguinal canal near the middle of its posterior wall. To accomplish this, it must either dilate a portion of the conjoined tendon into a sac covering, or split and separate its fibres from each other; or, in some cases, it may escape outside the conjoined tendon altogether, between it and the epigastric artery, pushing before it only the fascia transversalis. In the first case, which is most common, the sac is covered by an attenuated dilatation of the outer part of the conjoined tendon, rendering the complete investments of the hernia rather thicker and more resisting than in the latter. In the second case, the coverings are thin, but the edges of the internal opening more acute, and more apt to form a dangerous strangulation. In the last case, there may be a close resemblance in appearance to an oblique hernia, even extending to a covering of a few of the fibres of the cremaster muscle spread over the sac, while the conjoined tendon remains untouched, forming the chief part of the deeper wall of the hernial canal.

When a hernial protrusion is once fairly established in the inguinal canal, the action of the *rectus abdominis muscle* upon its posterior wall tends powerfully to aid in its dilatation. Enclosed in a sheath derived from the upper part of the conjoined tendons of the internal oblique and transversalis muscles, and more or less attached to the fascia transversalis at the thin border of its outer tendinous origin,—the backward and upward traction of this muscle upon the posterior and superior boundaries of the canal tends to open up the passage and to admit of the further progress of the hernia. At the same time, its backward pressure upon the intestines forces them outward upon the yielding aponeurotic structures in the groin. On a close inspection of the hernial passage, it will often be found that the chief yielding takes place at the superior and posterior parts of its walls, and that the cavity within the abdominal walls is much larger than the degree of bulging of the anterior wall indicates. If the aponeurosis of the external oblique be strong, the arciform and intercolumnar fibres resisting, the external ring small, and its pillars well bound together, a considerable dilatation, at the expense of the superior and posterior walls, may take place with but little external evidence of its presence.

This condition of an inguinal hernia, confined within the thickness of the abdominal walls, with a bulging of anterior wall or of the fascia covering the superficial ring, is termed *bubonocele*. (See Fig. 4.) In such a state the rupture may continue for a greater or less period without making any apparent advance in size, and may be retained by the pressure of a truss without entire reduction of its contents; which, consequently, are still subject to the danger of strangulation at the internal opening. In bubonoceles of long standing I have often found a large internal cavity, with a superficial ring but little dilated, and a very slight external tumour. When, in such cases, the superficial ring is at length dilated, and the tumour escapes into the scrotum, it assumes the appearance of a direct hernia, though it may traverse the entire inguinal canal. The influence of the rectus muscle upon the hernial passage and backward protrusion of the posterior wall, has an important bearing upon the radical cure of hernia, as showing the necessity of including this part in the action of the ligature, and thus procuring its adhesion to the anterior wall.

The sac of an *oblique* hernia is placed in front of, and to the outer side of the spermatic cord, and exerts its dilating influence primarily and chiefly upon the outer pillar of the outer ring. Invested with all the coverings of the cord, the deepest or infundibular fascia keeps it in close relation with the elements of this important structure.

The sac of a *direct* hernia is placed to the inner side of the cord and its infundibular and cremasteric coverings; and, while in the canal, is more distinct and distant from it than an oblique hernia. It exercises its dilating action chiefly upon the inner pillar and intercolumnar fascia. When emerging into the canal outside the conjoined tendon, it is apt to get quite behind the cord, as seen by Le Dran, Schmucker, and Blizard. An instance is given by Sir Astley

Cooper, of the elements of the cord, as well as the cremaster muscle, being spread over the surface of the sac; the vas deferens being separated, by the lateral distention, from the spermatic vessels. (*Op. cit. Plate* V., *Fig.* 5.) A somewhat similar case is recorded by Camper. The possibility of such a condition must be carefully borne in mind by the surgeon, in treating hernia.

That part of the sac of a hernia which is involved in the abdominal wall becomes the *neck of the sac* when the tumour has passed the superficial ring, and entered the scrotum. The distention in the latter place is then termed the *fundus* of the tumour. In the *direct* variety the neck of the sac is short, and the opposing influence of the valve action of the canal is more or less completely destroyed. This renders it more difficult to produce an extensive union of its sides by the radical cure. It is, therefore, more liable to a return after operation than in the oblique variety, in which the neck of the sac is more extensively embraced by the canal, and in which the valve action of the latter may be completely restored, and a broader surface of adhesion secured. At the same time, the greater separation from the spermatic cord in direct hernia gives a greater facility for the application of ligatures to unite its borders to each other and to the coverings of the superficial ring.

The form assumed by the *external ring* under the distending force of a hernia of either kind is usually an obliquely ovate. The sides are formed by the pillars of the ring, the upper end by the intercolumnar and arciform fibres, and the lower by the crest and spine of the pubis. Upon the number, extent, and strength of the intercolumnar fibres depends the degree of resistance afforded by the superficial ring to dilatation, which is effected by a separation of the sides of the opening in the tendinous fibres of the external oblique. In open conditions of the superficial ring, the intercolumnar fibres are often arrayed in one mass or bundle placed above the middle of the opening.

Occasionally, in *oblique* hernia, the long diameter is so great that the opening feels somewhat like a slit. These constitute generally more favourable cases for operation. Sometimes the orifice is so widened that the long diameter is rendered almost transverse. In *direct* hernia the shape of both the deeper and more superficial opening has a tendency to be nearly or quite circular. This difference is accounted for by the more direct pressure from within, in the latter variety. The pillars form a much more resisting and compact tissue than the intercolumnar fascia, and are simply pushed aside and everted by the pressure of the tumour, while the latter is distended and dilated into a covering for it as it passes into the scrotum. In the *oblique* variety the outer pillar is usually the more everted, and placed at its insertion more completely behind the sac than in direct. Between it and the sac lies the spermatic cord, usually plainly distinguishable just outside the pubic spine. The intercolumnar arciform fibres, in large cases, are stretched and frayed out to the utmost above the neck of the sac. In a *direct* hernia the protrusion affects more and earlier the internal pillar of the ring, but, when large, the outer pillar also is compressed and everted; the spermatic cord is also pushed outwards so as to lie further from the pubic spine than its normal position.

It is of the greatest importance to recognise the eversion of the pillars of the outer ring just alluded to, in performing the operations for the radical cure hereinafter described; and to place the ligatures or pins so as to include the everted portions, and to bring them fairly in apposition by the pressure of these instruments.

Both the varieties of inguinal hernia receive an additional covering of the *intercolumnar* or *external spermatic* fascia, as they traverse the external ring. This fascia is often very strong in old herniæ, and, when folded, compressed and consolidated by the action of the ligature after an operation, forms an addition to the barrier against a fresh protrusion. In an *oblique* hernia, covered by the cremasteric fascia, that structure forms the thickest of the deeper coverings, and constitutes the greater bulk of the invaginated tissue after operation. As a rule, the coverings of a direct hernia are thinner and less bulky; many exceptions, are, however, found in old cases.

In some cases of scrotal hernia I have met with, the external ring has been so small, and its pillars so resisting, as to admit with difficulty the fore finger. The tumour in the scrotum, in these cases, was considerable, and the cavity within the abdominal layers so large, as to give a distinct hour-glass shape to the sac of the hernia (see Fig. 6). In the adjoined figure, taken from a dissection of an oblique rupture, the upper dilatation is so large as to distend and fray the thinner intercolumnar fascia above a centrally placed bundle of arciform fibres, and give a distinct double-bellied appearance to the protrusion externally. The firm resistance of the outer ring to distension, in such cases, seems to depend mainly upon the bulk and strength of this bundle of arciform fibres. The hernia had probably existed for a long time as a bubonocele before it finally burst through the ring into the scrotum.

In some cases recorded by Cooper, and in one or two met with by myself, a rupture has escaped from the inguinal canal, not through the superficial ring, but by splitting the fibres of the external pillar itself. In these instances, apparently, the strength and resistance of the arciform fibres has been concentrated in one spot lower down than usual, and the rupture has burst through or distended the thinner fascia binding the fibres of the pillar together above. In such a case the sac of the hernia, covered only by the infundibular and cremasteric investments, would lie close to Poupart's ligament, and immediately under the superficial layers of fascia. Thus it may simulate a crural hernia by resting more or less upon the adductor muscles of the thigh. In one of my cases, as seen in the Appendix (*Case* 22), this occurred after operation

FIG. 6.

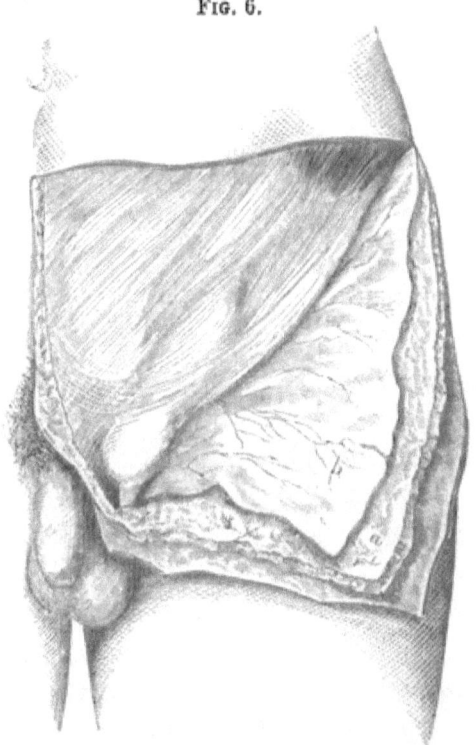

for the radical cure, through the internal pillar; probably through one of the needle punctures.

The principal thickness of the covering of a hernial sac is the superficial fascia, especially if the patient has a tendency to obesity. Its loose connexions, yielding nature, and the varying amount and want of permanence of the fat, which is its chief constituent in bulk, render it little to be depended upon as a barrier against the descent of a hernia; and in the operations about to be described, it is little used as an effective agent of security. Its looseness is so great in some patients that an inexperienced hand may invaginate the scrotum under the skin of the groin, so as to produce the appearance of having entered the superficial ring, without getting into the inguinal canal at all. Operations have even been done under these circumstances with a result that will easily be imagined.

As a general rule, the structures described in treating of the *anatomy* of *inguinal hernia* are demonstrated more decidedly, and are more evident to the touch, when a hernia has been present in the canal for any length of time; and the looseness of their connexion to each other in such cases permits of a more distinct discrimination. This is the case to a remarkable extent in old scrotal herniæ.

The different points at which the layers of the abdominal walls react upon a distended hernial canal and produce *strangulation*, gives us some hints as to the assistance which Nature is disposed to render in effecting a closure of the canal. The most frequent situation is at the internal opening of the sac into the peritoneal cavity. Except in a certain class of cases, the increase of danger consequent upon entering with instruments the abdominal cavity itself, will deter the surgeon from including freely this part of the rupture in his manipulations. He can, however, so operate as to cause adhesion of the parts which lie superficial to and around it, so as to cause them to become bound together, and prevent protrusion through the opening. The next point in the order of frequency is the lower border of the internal oblique and edge of the conjoined tendon. These structures, by the methods given in the succeeding pages, may be fairly and easily included in the ligatures. Occasionally, but least commonly, and generally in large ruptures, strangulation occurs at the external ring. This part of the canal is necessarily included in reaching the parts above it, and may be reached and operated on with great ease, and effectively closed by adhesive action. But if no more be effected, its closure does not prevent the formation of a bubonocele in the unclosed canal above, which remains sufficiently long to admit of a considerable knuckle of bowel, and to permit of strangulation at the inner opening. At the same time, the narrow line of adhesion accomplished by the operation will in time yield before the persevering efforts of the protruded bowel to escape into the scrotum.

In a recently formed hernia of either kind, in which the sac has not had time to form strong adhesions to the spermatic cord or canal, it may be detached by the finger, and pushed up into the abdominal cavity. The sides of the canal may then be drawn together and made adherent, without implicating the serous membrane at all. The elastic resiliency of the tissue composing the sac, when not diminished by too much or too long a state of distension, will speedily obliterate the "cul de sac" when the distending force is withdrawn. In such cases, the sac partaking more entirely of the character of peritoneum, and deriving its vascular supply altogether from its parent source, is, perhaps, less tolerant of operative interference than a sac of greater age. It may therefore be more satisfactory to some surgeons to avoid implicating it in the operation, if possible.

In ruptures of long standing, especially if scrotal, the sac contracts intimate vascular connexions with the cord and scrotal coverings. These vessels are supplied almost entirely from external and parietal sources, and the sac itself has thus become a structure almost independent of its source, the peritoneum. This is the case to a still greater extent, if a gristly ring of consolidated effusion be formed around the mouth of the sac at its junction with the peritoneum, as frequently happens, cutting off its vascular and nervous continuations, and rendering it a structure more or less distinct. That under such circumstances it may be cut off, or subject to inflammation, suppuration, or even mortification, without giving rise to any symptoms of peritonitis, is an observation made long ago by surgeons of experience.

In the scrotum, the adhesions of the sac are much more close and extensive behind, and about the spermatic cord, than to the more anterior structures, especially in oblique hernia.

When invaginated into the hernial canal above under these circumstances, it usually drags up the cord with it and elevates the testicle. The circular fold or "cul de sac," formed by invagination, is thus rendered necessarily deeper behind than in front of the invaginated part. (See Fig. 20.) This fact has an important bearing in estimating the causes of a return of the protrusion after any operation by invagination of the sac.

The *shape* and *outline* of the hernial sac and canal present many varieties, the recognition of which is of considerable importance, since the surgeon may often take advantage of their aid in effecting a radical cure.

In the dead subject, the shape of the sac may be obtained very readily by injecting plaster of Paris, and thus forming a cast of the interior. The outlines of the chief varieties I have met with are given in the margin.

Fig. 7.

As a rule, it will be found that the irregularities are in the inguinal canal, affecting the sac near its neck, precisely in that part most important to the operator.

In most cases of oblique hernia, the narrowest part of the sac is at the internal opening, and the sac is more or less pyriform (Fig. 7). In some cases of congenital hernia, and frequently in bubonoceles, the sac may continue of the same diameter throughout (Fig. 8). In many cases the sac is compressed or flattened antero-posteriorly between the layers of the abdominal wall.

In some old bubonoceles, the widest part of the sac is found at the internal ring, a large cavity existing in the dilated inguinal canal (Fig. 9).

In some, there is a well marked constriction at the lower border of the internal oblique muscle, extending more or less to the inner and posterior part by tension of the conjoined tendon (Fig. 10).

In many of the scrotal variety, as before alluded to, the superficial ring is more resisting than usual, and indents the sac all round (Fig. 11). The impression caused by unusually strong and compact arciform fibres in front of and above the neck of the sac has also been before mentioned (Fig. 12). So also has the occurrence of a double or many times repeated sac, one above the other, and separated by a perforated diaphragm of thickened and callous serous membrane (Fig. 13). In some cases this partition is more imperfect, a ridge or ledge being present on one side only. This is usually found at the inner and posterior side of the sac, being originally formed at the more acute part of the margin of the deep ring (Fig. 14).

In direct hernia, the shape of the sac is usually more simple, the constricted portion being at the opening in the con-

Fig. 8.

Fig. 9.

Fig. 10.

Fig. 11.
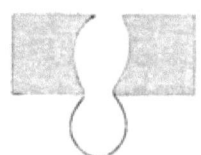

Fig. 12.

Fig. 13.

Fig. 14.

Fig. 15.

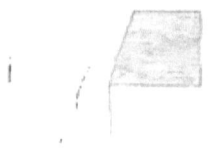

Fig. 16.

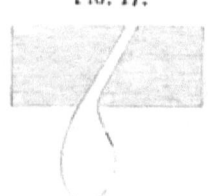

Fig. 17.

joined tendon; and the fundus, globular or flask-shaped under distension (Fig. 15).

When the hernia is fairly in the scrotum, the uniform yielding of its coverings usually gives a more or less globular shape to the fundus. To this there is an occasional exception in the presence of a slight hour-glass constriction in front, impressed by a more compact line of fascia about the middle of the scrotum (Fig. 16). This appearance is not uncommonly seen in hydrocele of the tunica vaginalis. In some cases, the scrotal tumour approaches closely in form to the pyriform shape of a hydrocele, with a small inguinal passage, especially in muscular subjects who have not worn a truss (Fig. 17).

I have myself met with a few cases in which there was a formation, more or less complete, of a double sac within the inguinal canal; one protruding at the internal ring, and another

Fig. 18.

through the conjoined tendon in the usual site of a direct hernia (Fig. 18). A case of such a kind will be found in the Appendix, cured by operation by ligature (*Case* 21).

Fig. 19.

Other cases are recorded, in which a double sac has been found protruding at one internal opening in the abdominal wall (Fig. 19). The explanation of such cases seems to be, that the neck of a small primary sac has become thickened and closely adherent to the fascia transversalis or conjoined tendon, so as to be immovable downward under the pressure of the intestines, while the loose serous membrane on one side of its neck has protruded under the pressure into another sac. A remarkable instance of direct hernia, where this occurred on both sides in the same patient, between the epigastric and obliterated hypogastric arteries, with the addition of a third protrusion on each side between the latter and rectus muscle, is illustrated by Sir A. Cooper, (*Op. cit.*, *Plate* VIII.) In this case the sacs passed through separate openings in the conjoined tendon, the interval between them being in both cases very small. Another more simple case is also figured by this author in the same work.

The presence of a tumour upon some part of the cord may modify considerably the shape of a hernial sac; as may also the co-existence of a hydrocele of the tunica vaginalis.

In children, its investments may be modified by the invagination downwards before the pressure of the fundus of a large tunica vaginalis; constituting the variety called *infantile hernia*. This variety, however, does not appear to call for any particular modification in the attempt to produce a cure.

In operating for the radical cure of hernia, a ready surgeon, following the principles recommended in the following pages, will take advantage of and employ most of the irregularities just described, in applying his instruments. The same peculiarities, however, will be found to present difficulties almost insuperable, in the treatment by invagination with a plug. So much has this been experienced, that various contrivances for increasing the internal dilating power of the instrument have been had recourse to, with the usual result of failure; being based upon the false principle of dilatation of the openings in order to effect their closure.

General Characteristics of the Subjects of Inguinal Hernia.— An extended observation of cases of inguinal hernia will not fail to recognise at least three different conditions of individual development co-existent therewith.

In *one class* of cases, the muscular system is well developed. The recti are distinctly defined, curving boldly at the outer border, and between the *lineæ tranversæ*; and so broad as to reach nearly half-way over Poupart's ligament. The external oblique is massy, and its muscular outline well defined. The internal oblique and transversalis hug closely the spermatic cord; the former muscle passing well down at its origin from Poupart's ligament, so as to cover the internal ring. The cremaster is powerful, affording a thick covering to the finger when passed into the hernial canal. The hips are narrow; the anterior iliac spines well closed in; the inguinal canal and Poupart's ligament short; the genital organs small and well braced up. The pillars of the superficial rings are usually tense, and easily recognised by the finger. The outer pillar is sometimes narrow and cordlike, and for some distance from the pubic spine is completely identified with Poupart's ligament. The distinction between the margins of the ring and the coverings of the hernial sac is very evident to the touch; the tenuity of the intercolumnar fascia contrasting with the firm and tense pillars of the ring. The arciform fibres often form a sharp impression across the neck of the sac. The hernia is frequently direct, sometimes separating the fibres of the outer pillar. The sac is small in diameter; but, when scrotal, much elongated, with a narrow and flask-shaped neck; the fundus apt to become pyriform, and, if irreducible, to simulate a hydrocele, or other affection of the cord and testicle.

Under the dissector's knife, the tendinous structures are found to be firm and strong, but thin and disposed to separate in fissures, from the delicacy of the connecting tissue. The fasciæ are fine, but tough, elastic, and resisting, containing but little fat in the meshes. The peritoneum is pliable, elastic, and translucent; and the whole of the abdominal layers are flexible, but glide upon each other to a limited extent only. The tissue of the sac itself is thin; the areolar network of its fibres spread out and frayed, still, however, retaining a considerable portion of elasticity.

The subjects of the rupture are able-bodied men, soldiers, sailors, or labourers. The rupture occurs suddenly, with a sense of something giving way, often with pain and sickness; in consequence of violent simultaneous action of the powerful abdominal muscles, compressing the viscera upon the comparatively thin tendinous and fascial structures. There is a want of proportionate development between the muscular compressing forces and the fibrous resisting tissues; resulting in a splitting of the latter, or a dilatation of the more yielding portions. These cases are prone to that violent form of strangulation mostly dependent upon spasmodic contraction of the internal oblique across the neck of the sac.

A *second class* of cases is marked by a greater development of the fibrous and fascial structures, with less extent and power of muscle. The subjects are wiry and sinewy men, often of a loose and shambling gait. The pelvis is large, with widespreading wings; the inguinal canal and Poupart's ligament consequently long, giving a broad appearance to the flanks. The tendon of the rectus is long, and the belly of the muscle begins to spread out at a higher point than in the first class of cases, leaving a greater space in the groin to be closed in by aponeurotic structure. The muscular part of the external oblique is small in proportion to the extent and thickness of its aponeurotic expansion. The internal oblique and transversalis arise from a small part only of Poupart's ligament. Their lower fibres, though tolerably thick, do not close up to, nor cover perfectly the internal ring. The cremaster is often large, and arises high up from Poupart's ligament, often so as to cover the internal ring, and thus to supplement, though less effectively, the imperfect downward development of the internal oblique. The total effect of this arrangement of the muscles and aponeurotic layers is to depress the belly towards the median groove, and to produce extensive bulgy projections towards the iliac wings and Poupart's ligament; an appearance which is so remarkable as to be recognised at a single glance, and indicates a hernial predisposition when no hernia is actually present. The genitals are usually large and loose, with a pendulous scrotum; indicating an abundance of fascial formation and a feeble development of the muscular fibres of the *dartos.*

The tendons and fascia are thick, close, and bulky; but inelastic, and apt to be weakened by rheumatic changes or fatty degeneration, especially after middle age, rendering them less able to resist pressure. The abdominal rings are long and open, and the canal loosely built. The pillars of the superficial ring are stout and thick; but lose themselves more imperceptibly in the thick coverings of the hernia sac, so that their edges are less evident to the touch. The outer pillar is often loose, baggy, and much everted, giving great play to the rupture, or to the surgeon's finger placed in the inguinal canal. The hernia is usually of the oblique variety, and of slow formation, increasing by starts after greater exertion than usual. It is often large, with a wide and infundibular neck, and a large internal opening. Its investments are generally thick, firm, and resisting, especially where the *arciform* bands pass across the neck. When these are unusually developed, which is often found in these cases, rendering the superficial ring small, the rupture is apt to remain for a long time as a bubonocele; dilating into a large cavity the internal ring and inguinal canal, and producing a bulge between the aponeurotic fibres above the arciform bands. When in the scrotum, the tumour is often rendered more or less hour-glass shaped, by an obliquely transverse band of strong fibres developed in its superficial investments halfway down the cord.

In these cases, a thick gristly ring is apt to form in the substance of the sac at its neck, or strangulation may be caused by thickened bands of the deeper fascia. The serous membrane of the peritoneum and sac is found to be thick, and not unfrequently opaque; and, probably, in old cases undergoes fatty degeneration, rendering it yielding and inelastic. Under the microscope, its yellow fibrous element is found to be more scanty, and to curl less vigorously than normal, while the areolar meshes are loose and large.

This kind of rupture is a favourable one for effecting a radical cure. The fibrous structures being thick and resisting, hold well under a ligature; and a large amount of consolidation and adhesion is usually the result. Barriers are often partially set up by Nature in the progress of the case; and these, though imperfect, may often be taken advantage of and fortified by the surgeon.

In a *third class* of cases, the groins and spermatic apparatus, and, in a less degree, the whole system, retain many of the fœtal peculiarities. The individual is usually plump, fat, and round. The pelvis may be ill-formed through rickets; having a tendency to the ovate deformity, with an unsymmetrical widening of the ventral notch, and spreading of the iliac wings. The abdomen is protuberant from imperfect abdominal muscular development. Towards the pubis the integuments are often wrinkled transversely; the mons veneris being large, loose, arranged in transverse masses of fat, and scantily clothed with hair. The penis is small, usually with a redundant prepuce; and the scrotum is shallow, broad above, and tapering inferiorly into the rounded infantile form. The spermatic cord is short and thick, and often encumbered with fat. The inguinal rings are very patulous, lax, loose, and capable of easy dilatation. The pillars of the superficial ring are small, thin, and not so easily made out as in the former cases, being gradually and insensibly lost in the coverings of the sac. They are weak, easily torn and cut, and do not hold a ligature well. No impression is made on the sac by the arciform fibres. The finger placed behind the outer pillar recognises Poupart's ligament as the only decided salient structure, and thin, narrow, and sharp. The inguinal canal is short and wide; the internal opening large, and its edges not easily distinguished. The lower border of the internal oblique is feeble and indistinct; and the conjoined tendon can only be felt by making strong traction forward, so as to raise considerably the lower edge of the muscle. The cremaster is feebly developed, and its fascia partakes of the general character of tenuity.

The muscular tissues are found, on examination, to be ill-developed; the fibres with a tendency to spread out, and separated by large areolar intervals. The fascia and coverings of the hernia are thin, the bulk of these tissues being made up of fat. The interposition of masses of adipose tissue permits the different layers to slide over each other more readily and extensively.

The peritoneum is lax and loose, and plentifully bestowed upon the superior false ligament of the bladder, which rises more than usual out of the pelvis when distended, and is broader in proportion to its depth. The serous membrane of the sac is more apt to be the subject of complications and adhesions than in other cases; and more frequently contains omentum, in which fat is apt to be plentifully developed.

These cases are commonly congenital or infantile in origin. Retarded development has probably led to a late descent of the testis, and to an imperfect closure of the inguinal openings, followed closely by a protrusion of bowel or omentum. It may attain to a very large size almost imperceptibly, and with little or no pain. Some of the largest ruptures I have met with were of this kind. In one which has lately come under my observation the testicle had almost entirely wasted away from the pressure of the tumour. This kind of rupture is perhaps less subject to strangulation than the others, because of the laxity of the circumjacent parts and the size of the internal opening. The same looseness of tissue, however, favours the occurrence of burrowing of matter after operation for strangulation or the radical cure. A certain want of vigour in the constitution also renders ultimate success in the latter operation less certain than in other cases.

Coexistent with this condition of the soft parts, the pelvis in such subjects often presents an increased degree of obliquity upon the line of the spinal column, attended with an increase of the forward projection of the lumbar curve and sacral promontory. The effect of this arrangement is to diminish the containing capacity of the cavity for the viscera; and to cause the abdomen to be unusually projecting, and to slope rapidly and deeply into the hollow of the groin. The weight and bearing of the intestines are thus, in the erect and sitting postures, thrown mainly upon the lower part of the anterior abdominal walls; and the pressure of the abdominal muscles is exerted more directly upon the inguinal openings. When such a condition is accompanied by a tolerable size of the deep hernial aperture, there is great difficulty in fixing properly the bearings of a truss, so as to be effective in keeping up the rupture. In extreme cases, indeed, I have experienced the difficulty of retaining the rupture, during muscular exertion, by any kind of truss or support whatever. The increased obliquity of the pelvis normal to young children, frequently baffles the truss-maker in the retention of congenital cases of rupture. Its persistence in the adult period, in the class of cases under consideration, may be taken as a further instance of the prevalence of the fœtal type. I have, however, met with it, though less commonly, in cases which I should place in one of the two former classes of cases, giving rise to much difficulty in the efficient adjustment of trusses.

It is commonly in the last class that those cases may be placed, in which the development of masses of fat in the sub-serous tissues, and within the coverings of the cord, by their presence and pressure increase the tenuity of the areolar and fibrous tissues, and open the lax spermatic passages.

In dissecting these cases, the peritoneum is usually found thinned to the utmost extent by the gradually filling up and dilatation of its areolar meshes by fatty deposit. In the omentum, particularly, it may be entirely perforated in many places, so as to assume a cribriform appearance; while vast masses of fat fill up the folds. This takes place also, to a great extent, in the parietal portion of peritoneum, especially about the kidneys and inguinal region. The serous membrane is of great tenuity, but at the same time retains much of its elasticity.

If in such a state of things the fat be quickly absorbed by illness or starvation of the patient, the more temporary tissue, —the fat, becomes used up more quickly than the containing tissues are able to contract and follow it. The inguinal, and other hernial openings are left patulous, and the weakened and yielding peritoneum quickly contributes a thin sac. A rapidly forming hernia is the result; the looseness of the layers, and their easily-sliding connexions, quickly accommodating them to the change.

Congenital Hernia. (See Fig. 68.)—In this form of rupture the sac is formed of that process of the peritoneum which, being drawn down along with the testis by the *gubernaculum*, forms, under conditions of uninterrupted development, the separate shut sac of the tunica vaginalis. The passage of the testis through the inguinal canal usually takes place about the eighth month of intra-uterine life. Under conditions retarding the processes of development, however, this transit may be delayed until after birth; and may be observed to occur in some children at any period during the first few years. Cases are not uncommon in which this process is retarded until the period of puberty has passed. In such instances, the gland is almost always retained permanently within the abdomen by adhesions to the colon or parietes; and is usually more or less atrophied. Sometimes it is arrested in the inguinal canal itself, and is not unfrequently mistaken for a hernia. The impulse given to the gland on coughing, under these circumstances, results from the presence of a portion of intestine or omentum in a "cul de sac" of the peritoneum above the gland, arrested in its development into the tunica vaginalis and canal of Nuck.

There can be little doubt but that the main cause of the formation of congenital hernia resides in the retardation and want of vigour of the developmental changes which seal up the inguinal rings and canal, after the testis has accomplished its transit. If the gland at the period of birth has but just escaped from the canal, or still lies lodged above the superficial ring, the cries and struggles of the infant during its first powerful inspiratory efforts, will force down a portion of bowel through the deep opening into the canal. The habitual recurrence of this protrusion will afterwards prevent the due closure of the openings. The cure or permanence of the rupture will then depend upon the issue of the struggle between the power of developmental change and the habitude of displacement of the bowel. As a rule, the later the descent of the testis through the rings is accomplished, the larger and the less disposed to close is the rupture opening which succeeds.

In many individuals not ruptured in childhood, a late descent of the gland leaves a patulous condition of the superficial ring, which predisposes greatly to the subsequent formation of a hernia; the only resistance in such cases being a limited extent of adhesion at the upper part of the canal and deep opening. It is generally associated with the feebleness and deficiency of the lower fibres of the internal oblique muscle before alluded to. A similar formative deficiency in some of these cases, extending to the gubernacular contractile apparatus, may account also for the non-descent of the testis.

In the early period of life, while development is proceeding with more or less vigour, the tendency of an inguinal rupture, as in the case of an umbilical protrusion, is always to close up and become obliterated. To aid the process, and in most cases to render it at all effective, it is necessary to exclude the bowel completely from the sac for a sufficient length of time. This is not frequently accomplished by the ordinary way of applying truss pressure; and, in most cases, demands a more effective method. From the time that the sac was last dilated by the intestine, we must continue to date the commencement of the curative process.

The first part of the *canal of Nuck* to close, after the passage of the testis, is its communication with the peritoneum at the deep ring, leaving the cicatrix, before described, at that place. Occasionally, at this stage of the process, an extraordinary degree of pressure of the viscera above, will protrude a distinct sac of peritoneum through the still patulous opening in the fascia transversalis; and by involuting the upper part of the elongated tunica vaginalis in the manner of a double nightcap, will produce that form of rupture called "*Infantile Hernia.*" This is characterized by the intervention of three distinct layers of serous membrane, between the skin and that which covers the intestine. Of these, the two outer belong to the invaginated tunica vaginalis, and the inner constitutes the true sac of peritoneum. It will be evident that this kind differs in an important degree from the usual form of *congenital hernia*, in which, the sac being formed by the cavity of the tunica vaginalis itself, the bowel lies immediately in contact with the testis and its appendages.

It follows from the above description, that congenital rupture must necessarily be of the *oblique* variety in its relation to the deep ring and epigastric artery. In long-continued and neglected cases, the deep and superficial openings are often closely applied to each other, and the rupture becomes, in its treatment and the difficulty of its management, like a direct hernia. A true *direct* hernia is, in my experience, rarely seen in the child. One specimen, however, I have recently met with, and operated on successfully for the radical cure. (*Case* 58, *Appendix*.)

III.

DIAGNOSIS OF INGUINAL HERNIA.

In no other disease is a correct diagnosis more necessary, or a mistake more glaring, and in certain cases more dangerous, than in the treatment of this complaint; whether urgent, or palliative, or with the view of effecting a radical cure.

It is in the latter relation, especially, that it is proposed to consider its diagnosis in the present volume.

The complications resulting from its association with the spermatic cord and testicle, with their coverings, render it very liable to difficulty in diagnosis. It may be said that the side to which a mistake generally tends is, to that of treating a doubtful tumour as a rupture.

A hernial tumour, when small, is more or less globular, the most part of its surface being, in a greater or less degree, obscured and engaged in the abdominal wall. When large, it has a tendency to be flask-shaped; a distinct neck of uniform thickness fairly disappearing into and distending the abdominal opening, while the dilated globular fundus is lodged under the skin of the groin or scrotum. (See Fig. 4.)

The great characteristic of hernial protrusions of the abdomen, when not strangulated, is the peculiar distending impulse impressed upon them by coughing, sneezing, or other sudden effort of the respiratory muscles; giving rise to a quick compression of the abdominal viscera against the sides of the containing cavity. It is important carefully to distinguish the true hernial impulse from the general impulse of the abdominal walls which takes place in the above actions.

If a hand be placed upon each groin in a healthy subject, when a cough is given—it will be found that a *forward* impulse is communicated to the inguinal structures by the jerk of the viscera against them from behind. Now, if the finger be placed upon or within the superficial ring, a similar impulse will be felt, and in the same direction—viz., forwards, *outwards*, and a little downwards. This impulse is communicated by the posterior wall of the inguinal canal.

Again, if a tumour exist on the surface of the groin, or is connected with its aponeurotic structures, the impulse will be communicated to it in the same direction. Nothing is more common than to mistake this for a hernial impulse.

If an *oblique* inguinal rupture be felt under the same circumstances, the impulse will be felt to be directed forwards, and more or less *inwards;* and it will be a *dilating* impulse, similar in character to the dilating pulse of an aneurism, as distinguished from the pulsation impressed upon a tumour placed in mere contact with the artery. This dilatation, at the moment of succussion, may be rendered more evident by holding lightly upon the tumour the ends of a couple of pencils or quills, placed at a little distance from each other. They will be seen to *diverge* from each other at the projecting extremities; showing a distending influence in the basis on which they rest.

The value of the evidence derived from this symptom depends entirely upon a proper appreciation by the hand of the surgeon, of this dilating impulse. If the rupture be large, and have a narrow neck, the ballottement will be to an appreciable extent *consecutive* to the general abdominal cough impulse, from a retardation in the wave of impulsion. This retardation, however, is absent in hernial sacs with wide necks.

At an early stage of the protrusion, or even before a true hernial sac has formed, a yielding or bulge is observable at the moment of coughing, above the middle of Poupart's ligament; indicating a yielding condition of the internal ring, and a weakening of the deeper structures. I have found this condition present in a great many patients to whom no suspicion of rupture had occurred. It indicates a strong predisposition to hernia.

The degree of dilating impulse in a hernia depends much upon the nature of its contents. If it contain bowel (*enterocele*) it is more evident than if it contain omentum only (*epiplocele*). The fluid or gaseous contents of the bowel more readily transmit the pressure. If the omentum be thickened by inflammatory or other deposit, the impulse is still less marked. Adhesions of the contents of the sac to each other, or to the sac itself, especially near the neck, also diminish the impulse.

In certain cases, on the other hand, the dilating characteristic of the impulse is increased. This is remarkably the case when effusion has taken place into the sac itself. In a remarkable case of irreducible hernia, shown in the theatre of King's College Hospital two years ago, in a child sent to me by Mr. Colgate of Eastbourne, the presence of two hard bodies connected with the omentum, and which I considered to be peritoneal concretions of the ordinary albuminous nature and cartilaginous consistence, gave rise to a peculiar and distinct ballottement, by their contact with the hernial walls, while floating in effusion set up by their presence in the sac.

In large ruptures, the presence of scybalous or other hard substances, floating in the liquid contents of the bowel, will also increase the impulse communicated to the fingers.

It is usually greater when small intestine is contained in the sac than where the large bowel is present, on account of the greater liberty of movement allowed to the former within the peritoneal cavity, and its smaller calibre.

Another characteristic of hernial protrusions is recognised in the mode of their formation. They make their first appearance in the deeper parts, and grow from within and above, downwards and forwards, towards the most depending part of the superficial investments; and generally in the direction of the fascial or other structures constituting a natural outlet.

The patient is usually made conscious of the formation of the rupture by a sense of something giving way, or of a greater weakness on the affected side when under muscular exertion. In a certain proportion of cases, the sudden sensation of a rupture having just occurred is absent. When the increase of the tumour is rapid it is always attended by a certain degree of pain. This, as well as the size of the swelling, is increased by any disturbance of the functions of the intestinal canal. Flatulent distensions and dyspepsia, with constipation and occasional obstructions occur. In ruptures containing omentum only these signs are, however, absent or faint. In congenital hernia they are also less marked. If the patient becomes corpulent about middle age, from an accumulation of abdominal fat, omental hernia increases more rapidly in size.

The course of the tumour in inguinal hernia is wholly, or in part, through the spermatic canal; and its tendency is to increase towards the scrotum or labia majora.

Its growth is neither uninterruptedly gradual nor uniform. In its earlier stages, the tumour appears only when the patient is in the erect posture. It becomes larger on his making any extraordinary muscular exertion, and disappears altogether when he lies down.

Frequently it remains absent for a longer or shorter period, even when the patient is standing and moving about. This peculiarity seems to depend upon the state of distension or flaccidity, which the bowel immediately opposite to the opening of the sac happens to be in at the time. An empty convolution will pass into an opening which will not admit of one more distended. This total disappearance of the hernial tumour, continued for some time, even during moderate bodily exercise, has a favourable bearing upon the probability of a radical cure being effected, inasmuch as it indicates a small internal opening, or a narrow and elongated neck, with considerable efficiency still remaining in the valve-like action of the walls of the canal.

A hernial tumour gradually disappears under pressure properly applied by the *taxis*. The best way of applying this manual force in a hernia of tolerable size, is by firmly grasping the fundus of the tumour in one hand, and making steady *compression;* while the other kneads gently the neck or deeper part of the tumour between the finger and thumb, with a view to returning first that part of the contents which emerged the latest.

The subsidence of the tumour under the taxis is sometimes rapid, and is attended by a gurgling noise, if intestine be present. If omentum only be contained, the reduction is effected more slowly, gradually, and silently. In enterocele, the tumour is more round and elastic; in epiplocele, it is more irregular in outline, doughy, and heavier to the touch.

The greatest amount of difficulty likely to be experienced, is in the diagnosis of a rupture while it yet remains in the inguinal canal as a bubonocele. In such cases, the superficial ring will generally be found large enough to admit the point of the finger; and a certain downward and inward impulse on coughing will be felt by the finger in that situation. If the thumb of the same hand, or the fingers of the other, be now placed upon the anterior wall of the canal opposite to the internal opening, the effect of the cough impulse in dilating the tumour will be rendered evident.

When an inguinal hernia has become scrotal, it forms a rounded tumour in front of the cord, resting below on the testicle, and can be traced by the touch to enter and become lost in the inguinal canal. The spermatic cord can usually be distinctly felt behind the tumour, and the testicle forms a distinct and evident mass below it, a little towards the posterior part. The cases in which the elements of the cord are separated, and the whole obscured by the hernia, are exceptional. Their nature may be recognised by the hernial impulse and other signs.

When of long standing, a scrotal hernia is often associated with a testicle more or less atrophied. So often has this condition come under my observation, that I am inclined to regard it as a common result of rupture, especially when the pressure of a powerful convex truss has been necessary to retain the rupture.

Some degree of varicocele is also often found in old cases of hernia, the result of the same conditions of pressure from a truss, or from the hernia itself.

The morbid conditions of the spermatic apparatus which are likely to be mistaken for inguinal hernia are various. One of the most common is an *undescended testicle* lying in the inguinal canal, and held there by a band of adhesion to the colon or parietes. These cases mostly come under the surgeon's notice in young lads. A considerable impulse on coughing, an increase of protrusion on making a muscular effort, and a considerable subsidence of the tumour on lying down, would lead to much doubt if there were not also the valuable indication of the absence of the testicle from the same side of the scrotum. A want of development of the scrotum on the same side, giving a striking and characteristic want of symmetry in the part, is usually found when the gland has not yet reached its destination. Pressure upon the tumour in the groin gives to the patient the peculiar sickening sense of pressure on this gland, and is usually accompanied by a degree of tenderness resulting from its abnormal position; thus giving a ready token of the character of the swelling. It is, of course, of great importance that its true nature should be recognised, since the pressure of a truss in these cases is most injurious.

Fatty tumours in the inguinal canal are sometimes met with, and closely resemble an omental hernia. The origin of these growths is often found to be in the subserous fat of the peritoneum. In this case they are much impressed by the cough impulse, and are likely by their enlargement to produce dilatation of the upper part of the canal, so as to be followed by a true hernial protrusion. For this reason it cannot be wrong to treat these formations by a proper truss, with the double intention of producing absorption of the tumour by pressure, and of preventing the formation of a rupture.

Development of an adipose mass in the substance of the cord itself, I have met with several times in dissecting-room subjects. This is the more likely to be taken for an oblique hernia, as it lies commonly in front of the elements of the cord itself, and is invested by most of its more superficial coverings. It is less liable to be impressed with a cough impulse than the foregoing; and may be distinguished by a careful exploration over the internal ring, at which point no cough impulse will be observed. The superficial ring may be dilated by such a growth sufficiently to admit the point of the finger. With one finger placed here, and another over the internal ring, while the patient coughs, a proper conclusion may generally be arrived at. The pressure of a truss would always be beneficial in such a case, for the same reasons as in the condition last mentioned.

Cysts may be developed in connexion with the spermatic cord in the inguinal passage. Though less common in this situation than in the scrotum, they give rise to more doubt in the diagnosis. They are usually formed in connexion with the vas deferens, either by the occlusion and dilatation of a diverticulum so commonly found upon the duct, or by the growth of an independent acephalocyst. The contents of a cyst of the former origin will sometimes be found to contain spermatozoa. In the latter case they contain a clear liquid, holding in solution some of the salts commonly found in the blood, but with little or no albumen. A case is recorded by Sir A. Cooper of a steatomatous cyst, placed upon the cord about the external abdominal ring, which was treated by a truss in mistake for hernia. In making a diagnosis it must not be forgotten, that a true hernial sac may be cut off by adhesion from its continuity with the peritoneum, and converted into a cyst presenting many of the characters of those just described. It usually contains a serous fluid more albuminous in its composition than the foregoing. Cysts are usually distinctly defined in their outline, shifting readily and eluding pressure; they feel very hard and plastic from tension of their contents, and are more or less globular. They are to some extent affected by the cough impulse, according to their position in the canal, but they may not implicate the deep ring, and have not the dilating character of a true hernial protrusion.

From the hardness and mobility of most of these cysts, the pressure of a truss, instead of producing absorption, will prove injurious by pressing the cyst into the superficial or deep ring, and dilating them so much as to predispose to a true rupture. In doubtful persistent tumours placed deep in the canal, therefore, it is better to suspend treatment and to watch the case. If the growth of the tumour be gradual and uninterrupted, and no sign of rupture follows, a puncture with a grooved needle will render the case clear which may at first be doubtful. Cysts superficial to the canal may be decided upon at an earlier period, and more easily, and treated by the ordinary methods.

Diffused hydrocele of the spermatic cord is often more difficult to distinguish, inasmuch as it possesses one of the most important characters of a rupture in disappearing altogether on the patient lying down, by the fluid passing through the distended areolar meshes at the internal ring into the subperitoneal tissues. In the erect posture it quickly returns, but without the hesitation frequently shown by a true rupture, and gives a certain impulse on coughing.

If the swelling, however, as is generally the case, extends down the course of the cord into the scrotum; its narrow and elongated form; its close contact with and obscuration of the cord, its fluctuation; and the marks of pitting left on it by pressure with the tops of the fingers, will lead to its recognition. A case of this kind which came under my observation presented this difference from a hernia,—much greater pressure over the deep ring and canal was necessary to prevent the swelling from reappearing on the assumption of a standing posture after lying down.

Glandular tumours are less common in the inguinal than in the crural canal and rings, but are sometimes found in the former situation, closely resembling a bubonocele. Like cysts, they are more moveable, more defined at the upper and outer part, and harder than a hernia, and may have no connexion with the internal ring. The presence of some condition of the testis or cord accounting for them, of a scrofulous diathesis or syphilitic taint, especially when accompanied by glandular swellings elsewhere, will lead up to a correct diagnosis.

It may be remarked in this place, that disease affecting the more superficial parts of the penis and scrotum induces irritation and swelling of those inguinal glands which are placed superficial to Poupart's ligament, and may finally lead to the enlargement of the glands which lie in the *crural* canal, through which their ducts pass into the abdomen. Disease of the testis and cord, on the other hand, affects primarily the lymphatics in the inguinal canal, and not the superficial glands. Such disease is usually evident enough to show the nature of a tumour of this kind simulating a hernia.

Abscesses may form in the inguinal canal. These may be of glandular origin and confined to that region; or may pass into it in the progress of a suppuration in the course of the vas deferens or round ligament. In the former case, they may be sufficiently removed from the internal ring to be unaffected by a cough impulse; and, in the latter, there will usually be present distinct evidences of coexistent disease affecting the prostate, bladder, uterus, or ovaries. In the ultimate progress of such cases, some degree of fluctuation, œdema, or tenderness will be apparent in the canal, which will explain the condition of the parts.

The rules ordinarily laid down in the text-books on Surgery, founded mainly upon the observations of Cooper, Lawrence, and others, will be found sufficient to distinguish the ordinary diseases of the testicle and cord, and their coverings in the scrotum, which are apt to simulate a scrotal hernia. In this place only a very brief allusion is required.

Hydrocele of the tunica vaginalis, encysted or otherwise, can give rise to doubt only when sufficiently large to enter into and dilate the superficial ring. Their growth from below; the greater obscuration of the testis and cord; their translucency on the application of a candle flame; their greater weight and evident fluctuation; and the absence of pain and cough impulse, are the distinctive marks. The shape of a hydrocele of the tunica vaginalis is mentioned by some as being distinctively pyriform; this, however, is subject to so many modifications, both on one side and the other, as to be untrustworthy, except in combination with other signs. Many cases of encysted hydrocele present themselves to the surgeon, in which the testis has all the appearance given to it by a hernia. The cord, however, is usually lost in the tumour; the inguinal canal is normal; and the superficial ring closed, unless in very large cases. The cord in these cases can be felt to be implicated in the tumour; by moving the testicle from side to side in the scrotum, its movements will be found to be followed closely by the tumour.

In the following pages will be found the details of a case of effusion into a closed hernial sac, closely resembling a hydrocele in its fluctuation, translucency, and shape. To render the resemblance more complete, the testicle was also enlarged, as is often the case in hydrocele.

Young children are occasionally affected by an effusion into the scrotal offset of the peritoneum, which may closely simulate a congenital hernia. In such cases, however, the diameter and obliquity of the upper part of the tumour which is so characteristic of congenital hernia is not so marked, the dilatation of the inguinal canal not being, in the earlier stage, so decided. (See Fig. 68.) Sometimes, the connecting canal being patulous throughout, the fluid can be pressed up into the peritoneal cavity, and the tumour made to disappear entirely. A cough and cry impulse, dilating the tumour, is often plainly evident, completing an exact resemblance to rupture in these respects.

The deep ring is frequently, in these cases, so small as not to admit of bowel passing into the scrotum, and the case will get well on the application of pressure and evaporating lotions. If the effusion be so abundant as to dilate the internal opening, or if this be large from the first, a portion of bowel will pass into the cavity, and a congenital hernia will be the result.

Not unfrequently, the internal aperture is completely closed at the ring, and the tumour cannot be dispersed by pressure; though still retaining a considerable amount of cry impulse, from its close contiguity to the peritoneal cavity in the inguinal canal.

On account of the thinness and delicacy of the coverings, in an infant these tumours may readily be distinguished by their weight, translucency, and fluctuation, as well as by the absence of rumbling or gurgling noises and intestinal motions under pressure with the fingers.

Doubts suggested by the shape and appearance of a *hematocele* will usually be dispelled by attention to the history of the case.

In certain cases of *chronic enlargement of the testicle*, attended with a thickened condition of the spermatic cord, suspicion of a hernia may arise. The deposit in such cases is usually of a tuberculous or carcinomatous nature.

I have lately had the opportunity of examining this condition of the testicle and cord after death. The patient was a young man who died suddenly while under treatment for the enlargement. The deposit in the gland was of the usual tubercular kind, the thickening of the cord was tolerably uniform, and extended along the whole length of the inguinal canal, but presented at points a kind of knotty enlargement caused by an accumulation of the same nature as that in the gland itself. The glands in the neighbourhood of the external iliac vessels were also enlarged and suppurating, involving the coats of the vein in the inflammatory changes; which had connected a softened clot, adherent to its interior surface, with the suppurating mass outside. The case had a clear history; and moreover, the hard unyielding nature of the swelling on the cord and the absence of hernial impulse, distinctly marked its nature and dissipated all doubts as to its true character.

Varicocele is a condition almost always popularly mistaken for a rupture. To the surgeon the characteristic signs are usually familiar. Its feel compared to that of a bunch of worms enclosed in a bag, eluding the pressure of the fingers. Its usually affecting first the lower part of the cord; its speedy subsidence when the patient lies down; and its gradual return on rising to the erect posture, while the finger is firmly pressed upon the superficial ring, which is usually of the normal size; are signs, in ordinary cases, sufficient to distinguish this condition.

In a certain proportion of the more severe cases, however, the varicose condition may extend to the spermatic veins lying in the canal; which may not only dilate the inguinal passage, but may give a decided cough impulse to the whole swelling. This varicocele cough impulse also closely resembles the true hernial impulse, in having a distinctly dilating character, from the fluid nature and internal communications of its contents. In some such cases considerable doubts may arise, more especially as this condition is often associated with a hernia; and the feel of a piece of omentum has some of the characters common to varicose veins. By close attention, however, the tortuous course, tubular character, and elastic reaction of the dilating venous channels can be satisfactorily made out; and, taken with the characteristics of a varicocele just enumerated, will render the case clear.

When a rupture containing omentum only is in the canal, coexistent with the varix, the case may assume a very doubtful character. In such a case the omentum does not recede so quickly, as the distended veins become emptied on assuming the recumbent position; and it may be recognised as a distinct structure by the point of the finger in the canal remaining still affected by the cough impulse. It will be found not to be variable in its volume, and not to afford the elastic reaction against pressure that varicose veins do. An inflammatory thickening in the canal, such as is common to varicose veins when injured, will be distinguished by the accompanying tenderness on pressure, pain, and œdema; affecting, to a greater or less degree, the testicle itself. By careful observation at different intervals, the presence of a portion of bowel, now and then, in addition to the omentum, will enlighten the surgeon upon the case. In the Appendix will be found the details of a very large oblique hernia, associated with a large varicocele, both cured by the same operation. (*Case* 8.)

IV.

HISTORY OF THE TREATMENT OF INGUINAL RUPTURE WITH THE INTENTION OF PRODUCING A RADICAL CURE.

A BRIEF review of the different methods which have been tried for the cure of rupture, may throw some light upon the causes of the want of success, which has been almost universal; of the danger to life, which has been too often demonstrated; and upon the essential requisites for effecting the desired end with safety.

The older methods comprise many plans, the palpable inadequacy of which render, to our eyes, their extension, adoption, and recognition only to be accounted for by extreme ignorance on the one hand, and impudent charlatanry on the other. Some of these pretended remedies, by plasters and ointments (most frequently containing *elecampane* in some form), serve occasionally, even in the present day, the deft hand of roguery in drawing contributions from the pocket of gullibility. Other and more severe plans of cure, by *caustics* or the *actual cautery*, possessed some more plausible claims on the imperfectly educated medical men of two centuries ago; and though less acceptable to the poor pain-dreading patients, were yet more readily submitted to than any process which would bear upon its back the formidable label of *an operation*.

One plan,—of causing ulceration of the coverings of the rupture by the use of *oil of vitriol*, obtained so famous a repute in this country at the early part of last century, as to lead the government of the day to bestow knighthood upon an impudent quack; and to give him, in addition, the more substantial reward of 500*l.* a year, and 5000*l.* paid down for his secret nostrum.

These severer plans were speedily found to be followed by such serious consequences, and such imperfect or futile results, as to be entirely abandoned..

Excision of the sac and its coverings, and in young boys, of the testicle also, as practised by Celsus, must have led to many fatal results, by exposing the abdominal cavity and its contents to much rough manipulation and deleterious external influences; as well as the general system to a severe shock.

Another plan, described by Galen and Paulus Egineta, consisted in *ligature of the sac* at and below the superficial ring; the ligature embracing also the spermatic cord and skin, the testicle being at the same time removed or left to slough out. For centuries after the time of the last-named medical luminaries, the practice of some modification of this plan is stated by Dionis and Scultetus to be extensively in vogue by unqualified charlatans perambulating among the poor and ignorant populations of Italy, France, and Germany; achieving serious and extensive public injury, and bringing into discredit all more rational plans for the cure of rupture. Such are the marked examples of the value placed upon a radical cure of this common complaint in all ages and countries.

As a competitor for public approbation and patronage against this wholesale mutilation of the king's lieges and military recruits, another, a more mild and scientific proceeding, was introduced under the prepossessing name of *the royal stitch.* This consisted merely in opening the sac freely and stitching the edges close by a continued suture, with a view to obliterate the cavity.

About the same date a plan, called the *punctum aureum,* was also practised by the more scientific and regularly educated medical men. As described by the celebrated Ambrose Paré, it consisted in passing a golden wire, or sometimes a leaden one was preferred, behind both the sac and spermatic cord at the superficial ring, and including both these and an intervening portion of skin in a loop; which was then twisted down tight enough to close the hernial opening, without totally interrupting the passage of blood to and from the testicle.

Of both these plans it may be said that, even if effectual in closing the sac at the point operated on, the result was simply the temporary conversion of a scrotal hernia into a bubonocele; with the certainty of a further protrusion at some future period, sooner or later. The closure of the inguinal canal itself was not attempted, and yet the exposure of the sac, and through it the peritoneum, to the air, would render them liable to the bad consequences which have at times ensued from this cause. Of the last-mentioned plan it may be said further, that there was every likelihood of troublesome complications arising in the testicle, if the ligatures were twisted a little too tight to allow for the succeeding tumefaction; and of direct failure if not twisted tight enough to close the hernial channel.

The operations practised at a later date by Schmucker and Langenbeck of Berlin, consisted in exposing the sac by a *free incision* at the superficial ring, separating it from the cord, and passing a ligature round the sac alone. The former surgeon then proceeded to remove the whole of the fundus of the sac below; while the latter was contented with suffering it to remain in the scrotum, either to slough or to become obliterated by inflammatory adhesion.

The large proportion of fatal cases (two or three in ten), and the great severity of the symptoms in all cases resulting from these methods, together with the want of a successful result in many; as shown by the experience and testimony of these surgeons themselves, as well as of Acrel, Arnaud, and Petit, in

France, and of Abernethy and Sharp in this country; were such as to deter these celebrated surgeons from prosecuting their attempts, and the methods above described fell into disuse. It is evidently the extensive exposure of the serous membrane; the rough handling of the cord; and complete strangulation of the sac in these operations, which exposed the patients afterwards to the very severe symptoms and great risk of life recorded by the several operators; from primary peritonitis, burrowing of matter, and other mischievous consequences to the testis and cord. The theory assumed by Langenbeck,—that the neck of the sac from the point ligatured along the canal to the internal opening became degenerated and obliterated, like a tied artery, up to the nearest branch, was totally unsupported by observation; and from what experience has since taught us, will not, at this day, be received as even probable. Moreover, the tendinous openings being altogether untouched and uncompressed, a free course was permitted for matter to burrow upwards in the abdominal walls to the peritoneum; or if the patient escaped these consequences, a free exit was afterwards allowed for the return of the hernial protrusion.

The plan of producing a radical cure of rupture by the use of a *strong tight truss* with a hard pad of wood, was originally recommended by Richter. It has, of late years, been much practised by American surgeons, and by some in Ireland, upon the plan of Mr. L'Estrange. The results which have been thereby obtained have not been hitherto publicly presented in a satisfactory manner. The injurious effect upon the testicle and cord in some cases I have heard of, and the want of success in many others, have appeared to me to depend in a considerable measure upon a deficiency in the shape of the compressing pad, as will be described more completely in a proper section of the present volume.

It is in small and incipient cases only, that hope is to be

entertained of a radical cure from the pressure of a truss, however skilfully made and applied, so far as present experience seems to indicate.

The more modern operations that have been devised for the cure of hernia are so various, that some difficulty is experienced in classifying them so as to estimate fairly their different capabilities. Such a fertility of invention is an additional evidence of the activity of research and scientific ability which has been brought to bear, through so many centuries, and under such different conditions of anatomical and surgical knowledge, upon this important subject.

The methods most worthy of notice may be classed under two heads. *First*,—those which deal with the interior of the sac only, with a view of causing adhesion of its opposed surfaces to each other. This has been attempted in various ways—viz., by a simple seton of threads, candle-wick, or sponge, passed through the sac by a needle, or other instrument. Such are, briefly, the methods practised by Schuh of Vienna, and Riggs of New York. Belmas of Paris, passed into the interior of the sac a bag of goldbeater's skin, which was then distended by jelly, with the same object; proceeding upon the principle of the radical cure of hydrocele. M. Velpeau in Paris, and Professor Pancoast in New York, injected into the sac a solution of tincture of iodine or cantharides.

Of all these plans it may be said that, even if effectual in accomplishing their object without danger to the patient (which, for many of them, is a very liberal concession indeed), a simple adhesion of the surfaces of the sac itself, will not prevent the formation of another sac from the abundant and loose peritoneum at the internal opening; since the real, effective boundaries of the inguinal canal and rings are left as patulous as before. This was long ago pointed out by Mr. Lawrence in his able treatise *On Ruptures.*

In the treatment by *seton*, the profuse and continued suppuration which is induced exposes the patient to an increased risk of the occurrence of burrowing of the matter between the abdominal layers, or of pyæmia.

In the same class as the foregoing, may be placed the operation proposed by M. Bonnet of Lyons, which consisted in passing needles across the sac and coverings, and thus pinching them up, as they emerge from the superficial ring. This proceeding evidently deals only with the anterior part of the sac below the inguinal canal; leaving the upper and posterior portions, and the tendinous boundaries unaffected. The result was the temporary retention of the hernia in the canal, and an ultimate return to its former condition; with the inconvenient alteration of a constriction, or incomplete adhesion, opposite the point operated on, increasing the liability to strangulation. The ultimate condition of the cases operated on in this manner, as I have been informed by an eminent fellow-citizen of M. Bonnet, was invariably unsatisfactory.

The *second* class of operations,—are those which follow the method of invagination of the skin and fascia of the scrotum, to plug up the hernial canal. Upon this principle are founded two operations—one, practised by Signoroni and Gerdy, and followed considerably of late years by different surgeons in this country, with various unimportant modifications; and another, originated by Wurtzer of Bonn, and followed by Rothemund in Munich, Sigmund in Vienna, and by Spencer Wells, Redfern Davies, and others in this country.

In *Gerdy's method*,—the skin of the scrotum, containing some portion of the fundus of the hernial sac, is pushed up into the superficial ring by invagination upon the finger of the operator. A curved needle, armed with strong ligature thread, is then carried along the finger, and thrust through to the surface of the groin on each side of the point of the finger. The ligature is then tied up, so as to hold the invaginated sac and skin in their new position, till adhesion has taken place in the interior of the canal. Signoroni used a piece of catheter for invaginating; while Gerdy attempted further, to cause adhesion of the opposed surfaces of the hollow cone of skin by the removal of the cuticle by caustic ammonia, and by placing sutures at the mouth of the invagination,—a proceeding which was usually futile.

In *Wurtzer's method*,—there is substituted for the finger of the operator, a stout wooden plug (variously modified by his followers), which is intended to fill up the hernial canal and openings, and to stretch them so much as to set up adhesive inflammation on the serous surface, all round the invaginated sac. The plug is held in position by one or two needles, passing out at its extremity, through the anterior wall of the canal in the groin, and fixed externally to a grooved compress of wood. Between the plug and compress, the folds of the skin, fascia, and sac are then forcibly compressed by a screw; with the object of producing their adhesion to each other.

The essential difference between this plan and Gerdy's, consists in the retention of the plug, (usually about seven days,) until it is supposed to have set up adhesive action to a sufficient extent, by its dilating pressure upon the opposed serous surfaces and boundaries of the canal. The theory is, that this will take place to such an extent as to obliterate the neck of the sac, to prevent the unfolding of the invaginated skin, and so to plug up the hernial opening by a permanently inverted cone of skin and fascia.

In all the foregoing methods of producing a radical cure of hernia, it is intended to puncture the sac, or otherwise to cause it to become obliterated by inflammatory effusion. It may be assumed, therefore, that the danger of an extension of the inflammation to the peritoneum itself is nearly equal in all. In this respect, however, we must except the objectionable practice of injecting the sac with irritant fluids, where the unusual danger of an escape of some portion into the peritoneal cavity is obvious, and has in many cases proved fatal.

With this exception, the records of the numerous cases which have been operated on up to the present time, both in this country and abroad, clearly prove—that the danger of serious peritonitis is by no means great, nor such as to cause any apprehensions on the part of the surgeon or his patient.

Such fatality as has been placed on record, following all the operations just reviewed, has been, in most instances, owing to the burrowing of matter between the layers of peritoneum and deep-seated fascia.

When small punctures only are made in the integuments, and a seton or ligature traverses through them the deep-seated tissues; and more especially, when distending force is at the same time used; free suppurative action is set up in the deeper structures; and the pus, not being able freely to escape through the obstructed punctures, makes its way along the areolar intervals between the abdominal layers; or along the spermatic cord into the pelvis; or through the inner opening of the hernial sac into the peritoneum. In this way, and as a secondary effect, peritonitis ensues; or the patient may succumb to exhaustion induced by profuse discharges from an extensive suppurating surface. Such dangers are, of course, more imminent if the pressure of the invaginating instrument produces a deep-seated slough, which has sometimes followed the use of dilating plugs.

Unlike the older and more severe operations previously mentioned, the failings of these more gentle proceedings do not seem to lie in their danger or fatality, but in their inefficiency to accomplish a good and permanent cure in a sufficient proportion of the cases operated on, as to give to the patient good grounds for expecting a successful issue.

Most, if not all, of Gerdy's thirty-six cases of operation had a relapse of the hernia; and the operation subsided from public observation until resuscitated recently in various forms and disguises. Of the ultimate condition of cases operated on by Wurtzer's method, it is very difficult to arrive at sufficient satisfactory information. In all the cases seen by the author, or occurring in the practice of nearly all the surgeons known to him, these results have been entirely unsatisfactory. With very few exceptions, the rupture re-descended soon after the plug was withdrawn; or as soon as the constant use of a truss was discontinued.

A very general impression prevails among surgeons, both in this country, in France, Germany, and the United States, that none of the foregoing operations have given such a promise of satisfactory results as to bring them into general use.

A few successful cases among a great majority of failures will not counterbalance occasional fatal terminations. Still more objectionable, are operations which render the use of a truss afterwards painful or impossible, as sometimes occurs.

Fig. 20.

An endeavour to point out the causes of failure in these operations will put us in a better position to judge of means adopted to overcome them.

In Fig. 20 it will be seen that invagination of the hernial sac from the scrotum into the inguinal canal leaves, by its reduplication, a circular fold or "cul de sac;" which, from its closer attachment to the spermatic cord and posterior structures, extends much lower down behind than in front. In Gerdy's operation, and in those following the principle of his plan, no attempt whatever is made to close this posterior fold. Adhesion is simply sought and obtained at the anterior part of the inguinal canal, at the apex of the invagination at (*a*); or is futilely attempted on the inner surface of the invaginated cone. In Wurtzer's method, the theory is,—that the circular dilating pressure of the plug suffices to cause adhesion of the serous surfaces round the entire circumference of the neck of the sac, as well as consolidation of the hernial canal. This I believe to be an entirely mistaken assumption, and one not founded upon accurate observation. Except under an amount of pressure quite unbearable to the patient, and sufficient to give rise to sloughing of the compressed structures, such a result does not occur,—and for the following reason.

Between the two hard compressing surfaces and the serous surfaces intended to be affected, are many mobile layers of tissue, doubly folded by the invagination, and expressly arranged by Nature to escape injury, and ward off pressure from the spermatic cord; viz., skin, fat, fascia, subserous tissue, and tendinous aponeurosis. Effectively to overcome their conserving power by pressure alone, nothing less than very severe inflammation and sloughing is necessary. Such a result, produced by the wooden plug, gives the operation a character at once clumsy and severe.

Into the posterior fold of the reduplication of the sac, then, in the generality of cases where these operations have been performed, a portion of bowel or omentum in due time makes its way, along the posterior wall, and behind the adhesions intended to restrain it. Gradually and not less surely, the complete dilatation of the canal, temporarily diminished in diameter by the invagination, is fully re-established. The hollow cone of skin is the first to become obliterated by its elasticity of reaction; the slender adhesions become elongated by its downward traction; the hernial protrusion follows close, and completes the dilatation of the hernial passage and the distension of the sac to its former dimensions.

In the cases which I have examined after failure of Wurtzer's operation, the superficial ring and canal have been more than usually patulous. The effect of the plug is to dilate the superficial ring and inguinal passage as far as it extends, and to prevent any adhesion of the sides together. Consequently, when the hernia returns, a larger protrusion than before is permitted. It is here that the inherent vice of the principle upon which that operation is founded becomes more apparent. To plug up and distend a dilatable opening with the view of closing it, is evidently a mistaken proceeding.

The patulous condition of the superficial ring permits the weight of the testis and its coverings, and the elastic reaction of the stretched skin and resilient dartos of the scrotum, to exert their full effect in dragging down and stretching the slender adhesions to the anterior wall of the inguinal canal. The hollow cone of skin becomes thus quickly obliterated; and the dilatation of the posterior fold of the invaginated sac is accelerated by the backward traction of the rectus muscle upon the posterior wall of the canal. In front of the lower part of the returned rupture, are borne the cicatrices of the ruptured or elongated adhesions.

It will be seen, in the progress of such cases, that the mass of inflammatory exudation apparent in the earlier stages is only temporary in character, sufficing for a short time only to block up the passage and keep up the rupture. The parts sooner or later resume their normal flexibility from absorption of the mass of consolidation. Nothing less than a close and extended adhesion of the permanent structures forming the walls of the canal, will suffice to keep up its permanent closure. Such a result, it is evident, cannot be obtained by dilatation.

In ruptures having a narrow inguinal passage, the skin can be invaginated only a very short distance into the superficial ring; the plug effecting a mere indentation into it, with a limited area of adhesion around the needles. A bubonocele thus remains in the chief part of the canal, the dilating pressure of which from above soon suffices to overcome such slight resistance.

In the numerous cases in which a considerable constriction of the sac takes place at the superficial ring, with a much dilated canal forming a large cavity within the abdominal walls (as described in the former pages of the present volume), the plug cannot be made to act by filling up the interior; unless either forcible dilatation of the superficial ring be previously made to admit a larger plug, or the plug itself is made to expand after its passage through the ring is effected. To effect either of these processes is difficult to the operator, and very painful to the patient; and may give rise to such severe inflammation as to add much to the objectionable character of the whole proceeding.

V.

THE PRINCIPLES OF THE OPERATIONS PRACTISED BY THE AUTHOR.

To effect a radical cure of inguinal hernia in a satisfactory manner, it is necessary to adopt a plan to shut up permanently the inguinal openings and canal, by drawing together and producing adhesion of its tendinous walls as far up as the internal opening. If we succeed in doing so much as to restore the valve-like action of its oblique sides, we shall restore the part to its normal power of resistance to dilatation and hernial protrusion.

The most obvious way of doing this is by the application of sutures through the sides of the canal, in such a manner as to ensure its complete closure, both behind and in front of the spermatic cord, sufficiently close to the latter to procure its adhesion without interrupting its functions.

To accomplish this, it is necessary to obtain a hold upon the posterior and superior boundaries of the canal, so as to cause their firm and close adhesion to the anterior and inferior wall; or, in other words, to unite the conjoined tendon firmly to Poupart's ligament in front of and above the spermatic cord; so that these structures shall be made to hug the cord as closely as in their normal condition. This union tends directly to counteract the dilating influence of the rectus muscle upon the posterior wall of the canal. To ensure success, complete union must be established along the whole length of the inguinal canal, which will afford a broad extent of adhesion, able to resist the future efforts of the bowel to open up the passage. Slender points of adhesion, or simple closure of the pillars of the superficial ring, will not afford this requisite amount of resistance.

A careful study of the anatomical description in the preceding pages, especially if aided by an inspection of the inguinal canal in a dissected subject, will show that the conjoined tendon may be taken up by a needle without difficulty, if the forefinger be first passed along the canal and placed fairly behind the lower border of the internal oblique muscle, and then made to hook it forward towards the surface, so as to raise the tendon to which it is attached into salient relief.

It will be seen that the border of this tendon is placed but a short distance from the margin of the deep ring on the posterior wall; and that the epigastric vessels are at this point screened from injury by the fascia transversalis, as well as by the finger of the operator placed in front of them. If a subject affected with oblique inguinal hernia be selected for the demonstration, it will be seen, as before remarked, that these tendinous structures are more distinct than in a subject not so affected, especially if the rupture be large and of old standing. The laxity of the tissues in most hernial cases also enables the operator to pass the finger behind the edge of the conjoined tendon itself; and, in some cases, even to push it between the peritoneum and the back of that tendon, so as to feel the edge of the rectus muscle. The looseness of the subperitoneal fascia at that part, enables this to be done without any tearing of the tissues. A ligature properly placed through the conjoined tendon, and then drawn outwards and forwards, encroaches on the inner side of the deep opening of an oblique hernia, so as to diminish the size of this opening itself. In this manner is obtained as near an approach to the deep opening on this side, as can safely be procured without danger of wounding the epigastric vessels, and interfering with the peritoneal membrane itself.

On the outer side of the sac, towards Poupart's ligament, a much higher point can be safely reached with the needle; which may be made to puncture the aponeurosis of the external pillar directly opposite to the internal opening. When these two points are drawn together by tightening the ligature, the deep hernial opening is compressed both above and at the sides. The induration which follows the retention of the ligature suffices to close, more or less completely, this part of the neck of the sac.

The ligatures, placed as above described, will be found necessarily to include also the pillars of the superficial ring, and to lie for some distance along the sides of the inguinal canal, crossing from one side to the other, so as to close them up to each other.

It will be seen, that in the various methods originated and practised by the author for the cure of hernia, this novel principle of compression and closure of the tendinous sides of the hernial canal in its entire length, prevails throughout.

In this important particular, they differ entirely both from the older and more modern operations; all of which either deal with the sac almost solely, or rely upon the principle of dilatation or plugging of the canal.

They differ also from the more effective, but more severe and fatal ones of Schmucker, Langenbeck, and others, in being of an entirely subcutaneous character. The sac, if punctured at all, is pierced by a small and valvular opening only; and remains deep-seated and removed from atmospheric influences. The connexions between the sac itself and the layers of deep-seated tissues, are not broken down or interfered with, being traversed only by the needle and ligature. In many of the cases given in the Appendix, the sac has probably not even been punctured; but, being recent and mobile, has been pushed up before the invaginated fascia beyond the ligatured points, which have thus closed the canal external to the sac altogether. In young, small, and congenital cases, this has been effected without any transplantation of the scrotal structures; but by simple closure of the hernial canal by the use of pins. In larger and wider cases of longer standing, however, the distance separating the sides of the hernial channel, and the advisability of retaining as much as possible of the sac coverings within it, have induced the author to invaginate permanently the front and outer parts of these coverings; and to fix them in the canal as a basis to tie down the ligatures upon. Thus is produced, by the succeeding inflammatory and adhesive actions, a solid agglomeration of adherent and resisting tissues.

In direct hernia, the circumference of the short canal at the neck of the sac is closely embraced by the ligature nearly all round, so as entirely to shut it off from the peritoneal cavity. The pillars of the superficial ring are by the same process closed round the invaginated fundus, so as to compress and enclose it in a folded mass, and so to fortify the deeper adhesions against the protrusion of the bowel.

When invagination is practised, a small opening in the skin of the scrotum, sufficient to admit the forefinger, permits of the easy separation of a sufficient extent of the deeper sac coverings, to enable the invaginating finger to discriminate more closely and distinctly between the structures in the canal; to reach a higher point of invagination than can be effected with the skin; and to guard from injury the peritoneum, viscera, and deep vessels, during the passage of the needle. A direct and free escape is thus provided, also, for any discharge which may form in the progress of the case, at the most depending point of the ligature track; in addition to a puncture at the opposite end in the skin of the groin. In this manner, the possibility of the burrowing or pocketing of matter is rendered as small as may be. This incision in the scrotum gives, also, a much greater firmness, steadiness, and security of hold to the invaginating finger; and counteracts the tendency of the scrotal layers to slide off the tendinous structures during the passage of the instrument through them.

The means employed to hold the structures together in their new position until adhesion takes place have, for reasons to be hereafter given, been changed from ordinary ligature thread to stout wire or pins.

The method of procedure in fixing them has necessarily been varied, according to the requirements of the material used, and the peculiar exigencies of the case operated on.

In the first twenty cases given in the Appendix, stout hempen ligature thread was used to draw together the structures. This was chosen with the idea that its greater flexibility rendered it more fitting than wire to follow the needle in the somewhat delicate manœuvres required. As an additional precaution in large cases, the ligature was passed also across the canal low down, close to the external ring. The irritation set up by a ligature thread was, at first, thought to be favourable to the formation of solid effusion in the deep parts. In some of the cases operated on considerable suppuration was the result, and it was observed that the amount of solid matter effused, and of permanent adhesion, was by no means in proportion to the suppurative action; but that, on the contrary, in some cases in which but little pus was formed, a greater degree of consolidating action and contraction than usual ensued. In some measure this was due, no doubt, to the state of health and condition of the constitution of the patient at the time of the operation.

Hearing much of the non-irritating properties of wire sutures, about that time reintroduced to public notice by Dr. Simpson of Edinburgh; and having had, unfortunately, a case which terminated fatally from pyæmia three weeks after the operation for the radical cure (*Case* 20), the author was led to try ligatures made of metal.

The diminution of the amount of suppuration which thereupon followed was so marked, that, with some modifications in the manipulation of the wire which were found necessary after a few trials, the plan now practised in most cases by the author was adopted with great advantage and success.

Although it will be seen, in the remarks upon Case 20, that the amount of suppuration was not so great even in that patient as in some others, and that the patient was nearly convalescent when attacked by the symptoms; yet it was indisputable, that to lessen the suppurative action would be an important step gained, especially as the desired consolidating and adhesive action was not in proportionate ratio.

The operation by pins was suggested by the effects produced in the treatment of varicocele and varicose veins, by the use of the rectangular pins originated by the author. It was thought that in children with congenital herniæ having narrow elongated necks, it would be possible to close up the yielding and rapidly developing tissues by means simple and free from danger; so as to present great advantages over the slow and uncertain process of pressure by a truss.

The results obtained in thirteen cases have been, so far, so satisfactory in these respects, as well as in the permanency of the cure, that this plan may fairly claim a good place among the radical cures of rupture.

In the operation by ligature thread it will be seen, that an external compress is necessary, to tie down the threads upon and to compress the more superficial parts between the ligature and the surface. In that by wire, as latterly practised, this proceeding, from which most of the pain suffered by the patient arises, has been abolished; or only had recourse to in large and exceptional cases.

In most cases, the scrotal fascia and sac fundus are invaginated, and retained in the canal by the adhesions set up by the ligature. In the operation by pins, the sides of the canal are simply skewered, as it were, together; and no permanent invagination is effected.

In all, the main principle of securing the conjoined tendon with the posterior and superior boundaries of the canal, on the one hand; and the anterior and inferior walls up to Poupart's ligament, on the other; is adhered to.

In some cases which I have seen, inattention to the former important and more difficult part of the proceeding has resulted in a return of the rupture. In some of the failures which have occurred in the early part of my own experience, I have considered the unsatisfactory result to be owing to a want of success in securing a good hold of the conjoined tendon, and thus leaving too much room in the back part of the canal close to the cord. This has led to the practice of applying backward pressure by a bandage upon the wire loop, as recommended in the description of the operation.

VI.

THE OPERATION BY LIGATURE THREAD AND COMPRESS.

This operation was described by the author in the *Transactions of the Medico-Chirurgical Society for* 1860, vol. xliii.

The *instruments* used in performing it are, *First,*—a stout, unyielding needle, mounted in a strong handle, with a blade much curved near the shaft, and less so near the point; so as to lie conveniently in the curve of the forefinger. The point is blunt, with tapering, wedge-shaped shoulders; formed, not to cut, but to split its way through the tendons. Near the point, the concave surface is flat horizontally, so as to take up readily a layer of tissue; while the convex part is rounded off laterally, so as to slide easily along without danger of puncturing the finger of the operator. Towards the handle, the blade is thicker in the opposite diameter, and more sharply curved than near the point; giving it more resisting power, and a tendency to come short up to the surface when pushed far into the tissues. The handle is long and stout; giving a good leverage to the operator, and enabling him to control the point with greater facility. (See Fig. 21, *b.*)

For those who have not acquired facility in performing the last-mentioned manipulation, I have devised a needle of the same form, but provided with a sliding guard; which, projecting from the back surface of the point slightly beyond it, covers it until it is fairly planted in a satisfactory position for puncturing. By drawing with the forefinger a trigger placed under the handle, the guard is withdrawn and the point exposed to

ON INGUINAL HERNIA.

a sufficient extent to make the puncture with ease and safety. (c.) The eye of the needle should be well drilled and smoothly countersunk; and when used in the wire operation, should be slightly grooved towards the point, to lodge the hook of wire which holds it to the instrument.

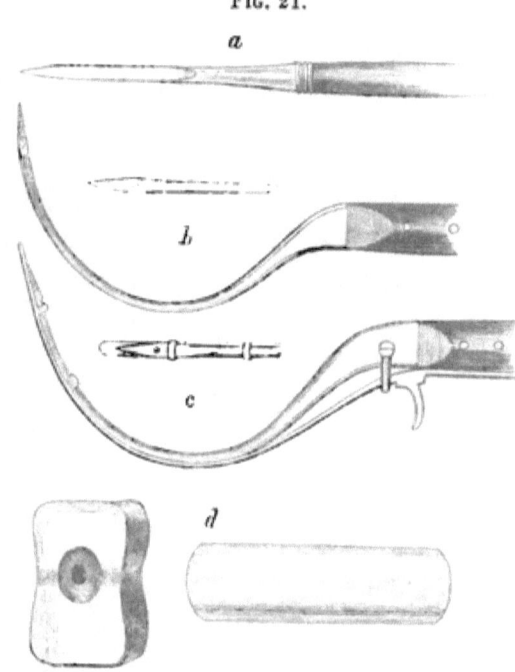

FIG. 21.

Second.—A small knife, with a blade resembling a tenotomy knife; the point sharp and lancet-shaped; the edge cutting only for an inch near the point; and a stout back. The handle should be thinned off at the end, so as to make a rounded edge like that of a paper-knife. This instrument is used for the purpose of making the preliminary scrotal incision, and separating the fascia from the skin. (*a.*)

Third.—A compress or pad, made of boxwood, glass, or porcelain, long and broad enough to cover the hernial canal.

OPERATION BY THREAD AND COMPRESS.

It may be cylindrical, flat, or boat-shaped, with a hole or groove about the middle to fasten the threads upon. (*d*.)

Fourth.—A stout hempen thread of sufficient length, free from irregularities, well waxed, and then soaped.

This operation is one in which much advantage is obtained by the operator from the administration of chloroform, which should be carried so far as to remove that spasmodic contraction of the abdominal muscles which involuntarily takes place under a painful manipulation of the deeper parts of the hernia, even when made for purposes of diagnosis only. In some patients this is difficult to effect with the usual expenditure of the agent. The stage of struggling should be completely passed before the needle is introduced. Violent contraction of the internal oblique muscle renders it difficult to obtain a satisfactory position for the finger and needle; while that of the recti and other muscles forces the rupture into the canal, and embarrasses the point of the invaginating finger at the critical time of passing the needle through the deeper structures. Accidents have occurred to the protruding bowel under the circumstances, which have mainly arisen, in my opinion, from a want of attention to the importance of passing the needle at the time of the most complete relaxation of the muscles of the patient under the anæsthetic agent. Great inconvenience also arises from cough, which may be set up by the use of an impure chloroform, containing irritating compounds.

The pubis and scrotum of the affected side being cleanly shaved, the patient is laid upon his back with the shoulders well raised. The surgeon will find it most convenient to stand on the side to be operated on, and to use for the purpose of invagination the forefinger of the right hand for the right side of the patient, and *vice versa*.

The rupture being carefully and completely reduced, the

finger is first passed into the canal by invaginating the scrotum pretty low down, to ensure a complete absence of the viscera from the sac, and to investigate the boundaries and peculiarities of the hernial passage. If the patient have a tendency to cough or struggle, or the hernia is easily protruded, an assistant must carefully command, by pressure, the internal opening during the preliminary incisions.

Fig. 22.

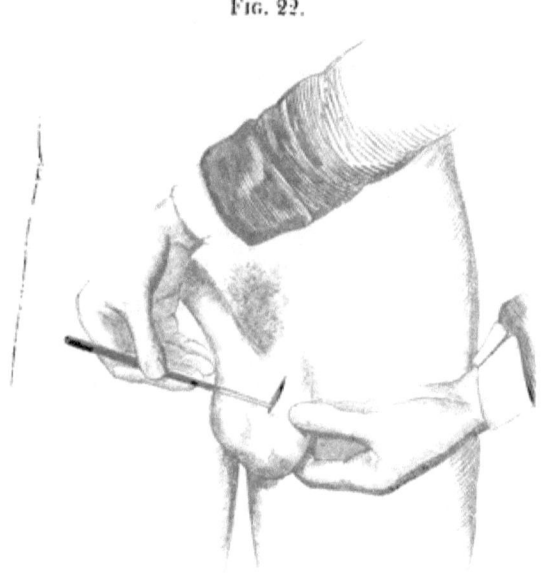

The Operation.—An incision is first made in the skin of the scrotum, over the fundus of the sac if the rupture be large, and a little below it if small. The most convenient direction of the incision for the future steps of the operation is obliquely downwards and outwards, terminating a little on the outer side of the scrotum. It should be long enough to admit easily the point of the finger with the needle in addition. (Fig. 22.) If the rupture operated on be a bubonocele, the point chosen for the scrotal incision should be $1\frac{1}{2}$ inches below the spine of the

OPERATION BY THREAD AND COMPRESS.

pubis. Then the knife, being insinuated flatwise between the skin and fascia for about an inch, is to be carried round the edges of the incision, so as to separate the former from the latter over an area of at least two inches in diameter. More than this will be required if the rupture be a very large one. The thin end of the handle of the knife will suffice to separate the loose connexions of the scrotal fascia to any extent that may be required, in ordinary cases.

Next, the knees of the patient should be brought together and bent up so as to relax the structures in the groin.

Fig. 23.

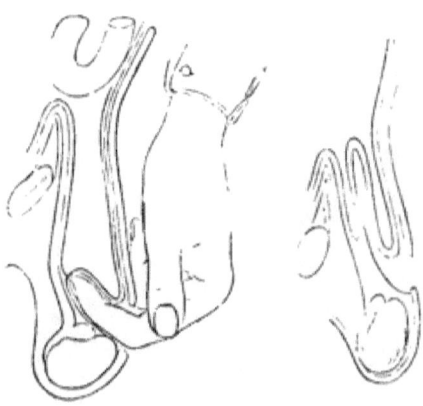

The operator's forefinger is then passed, with the nail directed backwards, into the scrotal aperture, and made to invaginate the detached fascia into the inguinal canal. This invagination should be commenced at as low a point as possible, so as to force the finger as much as may be behind the hernial sac, between its fundus and the spermatic cord. The latter may, at this time, be steadied by an assistant making gentle traction upon the testis. (Fig. 23.) The invaginating finger should be made to reach as high as possible in the canal, towards its superior opening. The position of the cord and of Poupart's ligament should then be distinctly made out. Then, by hooking forward the finger well towards the surface, the lower border of the internal oblique muscle will be felt raised upon it. This may be more distinctly recognised by placing the other hand upon the surface of the groin, when the thicker portion of the deep-seated structures in front of the rupture will be felt between the fingers. By directing the

FIG. 24.

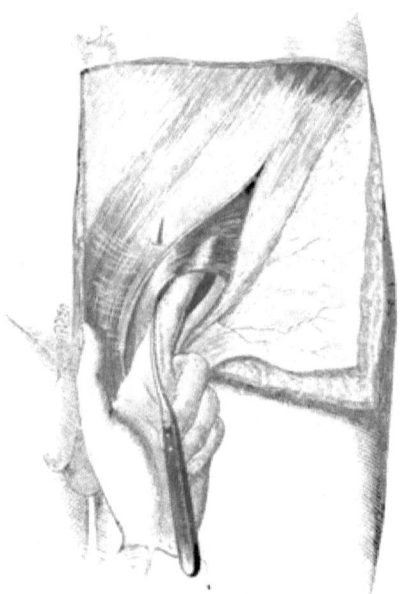

In this and many of the succeeding figures the skin and fasciæ are represented turned aside, to show the disposition of the invaginating finger and instruments, in the various steps of the operation. This takes place in the operation under these tissues, by the guidance of the sense of touch and out of sight of the operator, as seen in Fig. 25.

finger inwards, the operator will now feel at its thumb side the edge of the conjoined tendon raised with the muscle, and placed in relief on the posterior wall of the canal.

The needle, unarmed and well oiled, is then passed along the same side of the finger, and pushed through the tendon at its most salient part, so as to take up a considerable portion

OPERATION BY THREAD AND COMPRESS. 101

of it. (Fig. 24.) It is then turned towards the surface, traversing the internal pillar of the superficial ring obliquely upwards and inwards, till the point is seen to raise the skin of the groin. In these manœuvres the point of the needle should be carefully preceded and covered by that of the finger. The skin is then drawn inwards and a little upwards, as much as its deep attachments will allow, and the needle pushed through it. (Fig. 25.)

FIG. 25.

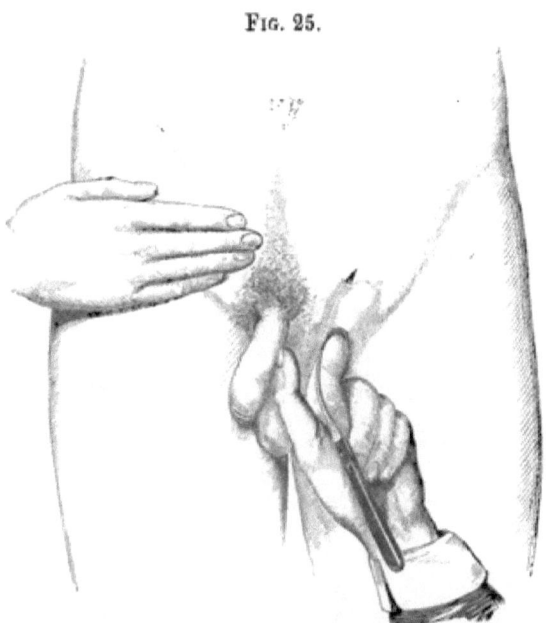

One end of the thread is then connected with the needle, and the latter withdrawn with a quick motion, leaving the other end in the puncture.

The invaginating finger is then placed behind the external pillar of the superficial ring, as close as possible to Poupart's ligament, opposite the internal hernial opening, in the groove which is there found between the spermatic cord and the ligament. The finger being again raised towards the surface, the aponeurosis is stretched well upon it. (Fig. 26.) The needle carrying the ligature is then passed along the finger between it and Poupart's ligament, and pushed through the latter opposite to the point of the former. When its point is seen to raise the skin, the latter is drawn outwards, until the needle can be pushed a second time through the puncture before made. A loop of the thread is then left in the puncture, and the needle withdrawn, carrying the free end.

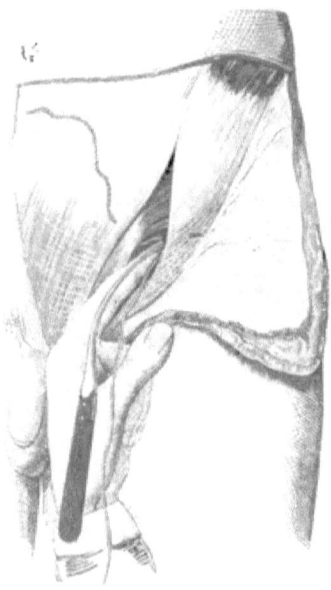

FIG. 26.

The finger is next placed on the inner side of the spermatic cord, just above the pubic spine, and pressed firmly upon the conjoined tendon, pushing it backwards and the cord outwards, so as to feel prominently the border of the rectus tendon. Into the tendinous layer of the *triangular aponeurosis* covering this part of the rectus, the needle is then thrust, so as to take up obliquely a considerable portion of that structure as near as possible to the pubic spine, which affords a good guide to the proper place for the puncture. (Fig. 27.) The point of the needle is then turned obliquely upwards towards the surface, and the skin drawn downwards and inwards over it, until it can be passed through the puncture for the third time. The needle is then freed altogether from the thread, and withdrawn. The whole of the ligature thread is now found in the upper puncture, presenting two ends and an intermediate loop. The upper end

FIG. 27.

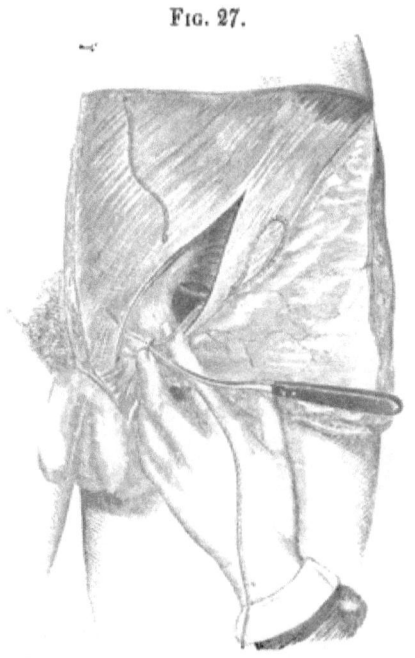

encloses the conjoined tendon and the internal pillar of the superficial ring; the loop passes through the outer pillar, close to the centre of Poupart's ligament; and the lower end through the triangular aponeurosis and the insertion of the internal pillar low down. Two portions of thread are thus placed across the hernial canal, invaginated fascia, and sac, closely embracing, but not including, the spermatic cord, and connecting the posterior or deep wall with the anterior or superficial (Fig. 28), perforating the aponeurosis in three places, as seen in A; but escaping by the same aperture in the skin, as seen in transverse section in B.

The compress, or pad, is then applied over the canal in an oblique direction, with its centre opposite to the threads as they emerge from the groin puncture. The two ends of the thread are drawn over to the outer side, and the loop to the inner, the latter crossing between the former. (Fig. 29, B.) One end of the thread is then passed over the compress and through the loop, and tied back to the other end in a loop knot or bunch (A). This method of fastening the thread in one instead of two portions gives an equable adjustment to the pressure.

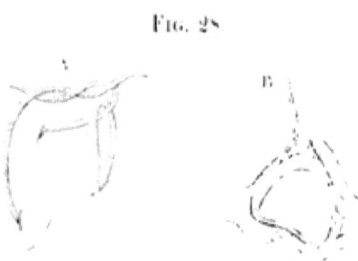

Fig. 28.

When the thread is tightened up, direct evidence of its action upon the canal and rings can be obtained by the finger. The posterior wall of the canal should be ascertained to be drawn forward by the ligature, and the pillars of the superficial ring closed in. If this effect on the posterior wall is not recognised, it may be apprehended that the conjoined tendon is not properly secured in the grasp of the ligature.

The lower end of the compress should reach as far down as the scrotal incision, which is usually tucked up close to the superficial ring by the traction of the ligature invaginating the fascia and sac. (Fig. 29, A.)

Pledgets of dry lint are then placed at the sides of the

OPERATION BY THREAD AND COMPRESS.

compress, and a fold of linen retained by a spica bandage over all.

The patient should be placed in bed with his shoulders well raised, and a bolster under his knees to keep the groin structures relaxed. If much pain be experienced, a full dose of opium in some form should be given to allay it, and repeated as often as this symptom may require it.

After the foregoing method were operated on the patients whose cases in the Appendix are numbered from 2 to 20.

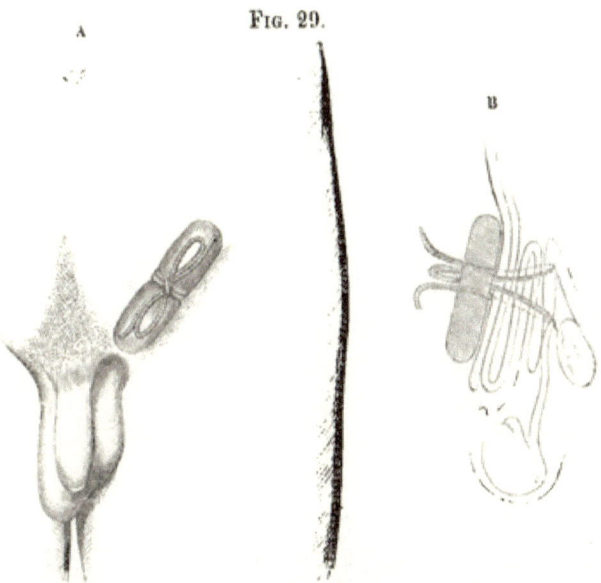

Fig. 29.

In the first case, the same principles of operating were also strictly followed, but two separate ligatures were used, and the instruments employed were of a more complicated character, which subsequent reflection and experience modified into the simplicity of form and number just described.

If the pressure were much complained of, the spica bandage was removed the next day. In small cases, or when much thickening and consolidation were quickly apparent about the canal, the ligature thread was untied, and the compress removed on the third or fourth day. In other cases it was kept on till the sixth or seventh day.

In some cases the discharge was small in quantity, appearing as a serous exudation on the second or third day. If the sac were freely punctured, the discharge was usually more abundant, in the form of a reddish serum, becoming purulent about the fourth or fifth day. Sometimes the lower opening in the scrotum healed entirely by the adhesive process. When the strain on the ligatures was great, and a large sac was included, the discharge escaped freely at both the openings. The threads were usually left in the wound to act as conductors of the discharge, and to keep up a seton-like action in the canal for as long a time as was deemed necessary, judging by the amount of solid effusion.

In one case (No. 8) in which a large rupture was complicated with an extensive varicocele, profuse suppuration in the interior of the sac ensued, and sloughs escaped freely by the lower aperture. The upper one discharged little; the ligature appeared to have cut off the sac entirely from the peritoneal cavity. No untoward complications whatever ensued, the patient made a most excellent cure, and was completely freed both from the rupture and the varicocele, while the testicle remained entirely uninjured.

VII.

OPERATION BY WIRE, AS AT PRESENT USUALLY PRACTISED BY THE AUTHOR.

The *instruments* used in this operation are the knife and needle before described, and a piece of *stout copper wire silvered*, about two feet long. The wire should be as thick as may be consistent with tolerable flexibility. The advantages of having the wire thick are—that a fine wire cuts the tendinous structures when drawn tight, acting like a tenotomy knife upon their tense fibres, so that these important tissues become set free from its grasp, instead of being held firm in their new position until made adherent by the slower process of ulceration. A thick wire also, when twisted, carries round with it, not only the included parts, but also the deeper structures attached to them; so that all the structures near the parts immediately traversed by the wire are dragged and twisted up more or less, so as to complete the entire closure of that part of the hernial passage which is not included in the grasp of the ligature. A thick wire also retains its form with more tenacity, and holds the openings which it traverses more in a line with each other; so that any discharge which may form escapes from the wire track with more facility, its passage being more direct and less obstructed. Copper wire is better than silver, because it is stronger, more tenacious, and not so liable to form kinks and break off;—it is better than iron, because at the proper thickness it is more flexible. Any disadvantages possessed by a thick wire are experienced at its bend in the eye of the needle.

If the doubled wire be as thick as the needle itself, the bend will not pass through the needle track with the point. The needle used in this operation, however, should be so thick as to be perfectly unyielding. Its broad blade will open a track sufficiently large to receive back with ease the point with the wire attached. Any little hitch will be readily overcome by withdrawing the needle with a rapid motion, accompanied by a slight jerk. In many instances, I have had the wire thinned down at the ends by gradations, made by passing them through the diminishing holes of the draw-plate for about six inches at each end. In some of the trials, however, the shoulders thus left on the wire were arrested in some part of the track, and the wire broke off—an awkward accident, necessitating a fresh application of the needle. Upon the whole, it is better to trust to a thick, stout needle with the eye grooved, and a wire of a degree of thickness limited only by convenient flexibility, and undiminished at the extremities. No. 20 is the degree of thickness I have generally used.

In preparing the wire for this operation, it should be drawn rapidly several times through a woollen cloth, so as to clean it perfectly, straighten it out, and by the heat of the friction to render it a little more flexible. About half an inch at each end of the wire should be carefully bent into a hook, sufficiently round to render its passage through the eye of the needle easy, and quite free from side twists. The wire should be then well oiled, folded in the middle, and given in charge of an assistant to keep it unaltered in form.

It has been remarked that the needle should be strong and unyielding. This is especially the case when wire is to be used, partly for the reasons above given, and partly because in performing the various passes of the needle considerable resistance by the muscles is sometimes experienced in drawing the tendon transfixed by the needle towards the proper place for puncturing the skin. In the figure given of the needle used by myself (Fig. 21, *a*), it will be seen that near the point it is flattened from side to side, but as it gets nearer the handle the diameter in the cross direction becomes the larger. This was found to be requisite to prevent the tendency of the curve to open out under pressure, and so allowing the point to slide between the abdominal layers, instead of coming short to the surface when intended to do so. Such an accident on the living subject might give rise to serious complications.

It will be noticed in the succeeding description, that in the operation by wire each pass of the needle is made with the instrument unarmed. With a wire of the necessary thickness, or with any kind of wire, or even thread, passed through the eye of the instrument, it is difficult, if not impossible, to transfix the tissues with the exactitude which is desirable. With the needle perfectly free, on the other hand, great precision is attainable by a moderate amount of practice.

The Operation.—The patient and surgeon being placed in the position described in the first operation, the same preliminaries being gone through, and the scrotal incision, the separation of the fascia and its invagination with the sac, being accomplished as before detailed, the border of the internal oblique is felt for and raised; the conjoined tendon recognised and carefully taken up on the needle, which is then made to traverse the more superficial parts obliquely upwards and inwards. The skin is then drawn in the same direction, and the needle pushed through it. So far the steps of the operation are precisely similar to that first described. (See Figs. 22, 24, and 25.) One end of the prepared wire is then hooked on to the eye of the needle, care being taken that it fits fairly into the groove, and the instrument is then withdrawn rapidly and with a slight jerk through the tissues, drawing the wire after it.

The needle is then disengaged, and passed upon the finger to its outer side as high up as the internal opening of the hernia, opposite which it is pushed through the anterior aponeurosis close to Poupart's ligament. It is then turned so as to traverse the same cutaneous aperture through which the wire has already passed, which is to be drawn upwards and outwards to meet it. The opposite end of the wire is then hooked on, drawn through the puncture after the needle into the scrotal aperture, and then disengaged as before. (See Fig. 26.)

Thus far the steps of the operation differ only in the one small particular of hooking on and off the ends of the wire, from the operation with threads.

At this stage of the operation the two ends of the wire emerge together at the lower or scrotal aperture, after traversing the conjoined tendon and internal pillar on the inner side and Poupart's ligament at the outer side respectively; while the loop which connects them emerges at the upper or groin puncture.

The sac of the hernia and the fascia covering it opposite the scrotal aperture is then pinched up between the finger and thumb, and the spermatic cord is slipped back from their grasp, in the same way as in the operation for tying the veins in varicocele. The needle is then passed from without inwards and a little upwards, in the direction of the incision across the scrotum, close to and in front of the spermatic cord. A slight twist given to the point of the needle will enable it to take up all the structures which lie in front of the spermatic cord, and at the same time to enter and emerge entirely within the limits of the scrotal incision. The curve of the blade and the extensibility of the scrotal tissues will permit this to be done easily, without making a fresh cutaneous puncture. If this be accidentally done, however, the incision may be extended so far as to meet the new puncture. One of the ends of the wire is then again hooked on to the needle (see Fig. 30), and drawn with it across the cord through or behind the sac, traversing the scrotal fascia. I have usually been in the habit of drawing that end of the wire which has already traversed the conjoined tendon through the scrotal portion of the sac, as being better placed, when drawn up, for making pressure upon the deeper parts of the canal, and more easily withdrawn afterwards. Either end, however, may be taken, the outer end having another recommendation of giving an extra twist to the sac when drawn up and tightened.

FIG. 30.

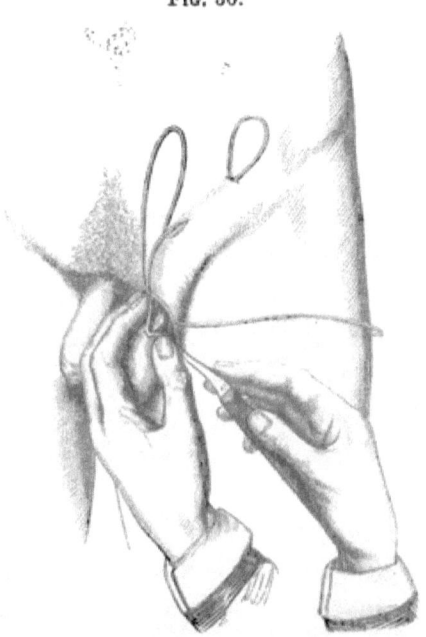

If the sac is small and recent, reaching only as far as the scrotal aperture, the wire may be thus placed entirely behind it, and between it and the cord. But if the sac be larger, and of long standing, its close adhesion to the cord will hardly permit this to be done, and it is unavoidably punctured posteriorly.

In still smaller cases, wherein the sac does not descend

much beyond the external ring, the last step of transfixing the fascia must be performed nearer to the insertion of the pillars of the ring.

In accomplishing this manœuvre, the needle may also be made to take up a portion of the pillars themselves close to their respective insertions on the inside of the spermatic cord. The crest of the pubis will afford a good guide and protection to the deeper parts, the point of the needle being made to slide close to the bone. This additional precaution is very desirable in large cases of inguinal hernia occurring in females, in whom there is not much fascia capable of being invaginated at this point.

In some very small cases, both in male and female subjects, this last transfixion of the fascia or pillars by the needle may be altogether dispensed with; as any great amount of permanent invagination of the sac or fascia is not necessary to fill up a narrow hernial canal when drawn together by the suture. In some small female cases, a simple incision over the superficial ring, without any separation of fascia, will suffice to apply the necessary sutures with accuracy, and to afford a free escape for the discharges.

The next step in the operation is to straighten, stretch, and draw down both ends of the wire, until the loop above is close to the skin. Here it is held by the finger of an assistant while the surgeon twists the ends round each other, giving them three or four turns. This manœuvre twists also the enclosed sac and fascia which are held between the ends of the wire.

Next, the loop is drawn steadily upwards so as to invaginate the twisted sac and scrotal fascia firmly into the hernial canal, stretching them as far up as possible towards its deep opening. The loop is then, in its turn, twisted well down into the upper or groin puncture, giving it the same number of turns. The forefinger of the operator should now be placed in the scrotal puncture to ascertain whether a satisfactory closure of the superficial ring has followed the tightening of the wire. The effect of traction on the wire upon the posterior wall can also be distinguished. Great care should also be taken that the skin of the scrotum is not drawn upwards between the pillars of the ring, so as to prevent their direct union.

Fig. 31.

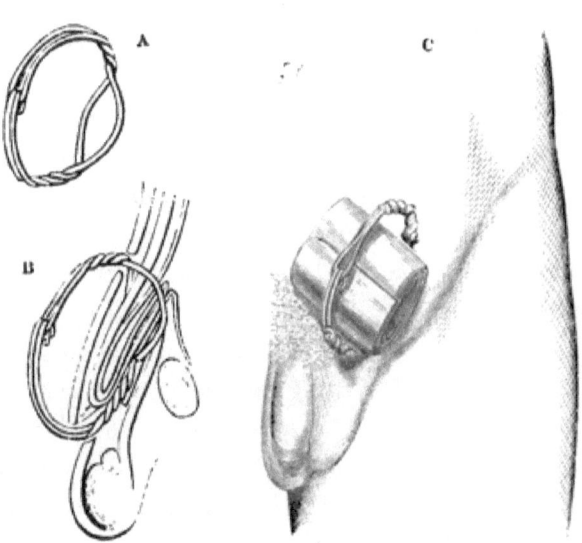

The projecting ends of the wire are then cut off by pliers about three inches from the surface; and both together bent into a hook, which is carried upwards to meet the loop curved down to receive it, till the two are locked together, and form an arch over the intervening skin. (Fig. 31, A, B.)

A pad of lint, rolled tightly up to a size sufficient to fill up the interval between the arch of wire and the skin, is then placed under it between the punctures. (Fig. 31, C.) A broad spica bandage is then placed over all, so as to make firm compression upon the wire steadied by the pad of lint.

In male cases, it is advisable to ascertain that the cord is moveable behind the lower twist of wire, and not included in its grasp. It should also be ascertained that the superficial ring is entirely closed up by the tightening of the wires.

In large scrotal herniæ, the testis on the side operated on will be more or less drawn up towards the ring.

When the patient is removed to bed, care should be taken to keep the abdominal structures relaxed, by raising the shoulders and placing the knees bent over a bolster. The scrotum may be supported by a small cushion covered with gutta percha skin. As a support to the testes, I have been in the habit of employing a long and broad strip of common adhesive plaster, carried across the upper parts of both thighs, and passing behind the scrotum. This is not so likely to permit the scrotum to slip down as the piece of wood used by Ricord as a support. Attention to this point will obviate any bad consequences in the way of swelling and pocketing of matter, which might follow a dragging and dependent condition of the testes.

A slight enlargement of the testis on the affected side, coming on in twenty-four or forty-eight hours after the operation, is a satisfactory indication, showing a sufficiently close embrace of the substance of the cord, and a sufficient pressure upon that part of the sac which immediately invests it, as well as upon the posterior wall of the canal. If this swelling be great, or there is much pain in the gland, the spica bandage may be removed after twenty-four hours' application. Generally, however, it will not be found necessary to remove it till the third day. General abdominal uneasiness may also necessitate its removal. To this, the application of a hot fomentation flannel will usually give immediate relief.

The after-treatment consists simply in giving opium in some form, as often as pain or restlessness renders it advisable. Patients differ much in this respect. For the most part, they suffer hardly any pain after the first twelve hours. In some cases they are restless, and complain of the tightness of the bandage, of flying pains in the abdomen, and of a flatulent condition of the bowels. No tenderness on pressure upon the abdomen is, however, experienced, pressure rather giving relief. Sometimes pain is complained of along the iliac crest, which I attribute to the pressure upon the small ilio-inguinal nerve, by the ligature enclosing it. Five grains of pil. sapon. co. given soon after the operation, and repeated every night at bedtime, or as often as rendered advisable by the pain, is the usual treatment I have pursued, with hot fomentations in some of the cases.

It is important in the after treatment of these cases not to keep the patient *too low*, as we thereby diminish the power of reparation on which so much of the permanency of the cure depends. The diet should be mild and unstimulating, and taken more freely when the patient feels inclined to eat it. To begin with a milk diet is best, with strong beef-tea ad libitum, and two eggs daily.

If the bowels have been properly cleared before the operation by a mild dose of castor oil, and the diet thus carefully regulated, it is better to leave them to their normal action afterwards than to irritate them by purgatives. If moved within the first three or four days it will be sufficient. If not, and the patient complains of distension, a tablespoonful of castor oil will be found to make him more comfortable.

After the bowels have been opened, the patient is generally allowed to have a slice of meat or a chop, if he can eat it; but no vegetables, until a longer time has elapsed, and the wound is fairly healing. If the latter process be somewhat sluggish, six ounces of wine daily will be beneficial. If the patient have been previously accustomed to stimuli, this may be increased to eight or ten ounces.

The wire usually gives rise to very little irritation or suppuration, and may be kept "in situ" much longer than threads, which seem by their retention of putrid or decomposing matter to cause more suppuration in the wound.

In the course of the first week, when the tumefaction of the more superficial parts has subsided, and the discharge, after being serous, has become purulent, evidences of thickening and consolidation of the transplanted tissues within the canal become distinct.

The effective effusion in the canal may be distinguished from an exudation into the superficial tissues, by the skin being moveable over it as freely as over the surrounding parts. In a short time, the effusion becomes so large and hard, that it may be held between the finger and thumb, and felt to be deep-seated behind the aponeurosis of the external oblique.

I have generally untwisted the wire about the eighth or tenth day, and removed it altogether about the fourteenth. In cases where the amount of solid effusion seemed to be less than usual and slow to form, I have kept it in as long as twenty-one days with benefit. Its withdrawal has rarely given much trouble, and but a small amount of pain. By acquiring the habit of giving the twist to the wire always in one direction, and counting the turns, it becomes easy to untwist in the opposite direction, and with the same number of turns reversed. The track in which it lies in the tissues is usually so much enlarged by ulceration as to permit of the wire to pass with slight traction, although one or two of the irregularities caused by the twisting may remain on it. The two wire tracks are usually, at this time, blended into one by ulceration of the intervening parts, and the wires may be felt rubbing against each other in their whole length. In large cases, where they are widely separated, this may not be the case. The movements of the two ends on each other will indicate their freedom from each other's embrace. It has usually been necessary to untwist only the lower ends of the wire; the loop and the upper twist generally coming away unaltered by upward traction. It is better always to draw the wires out upwards, if possible, to avoid the chance of tearing downwards the new adhesions of the invaginated fascia. In one or two cases only, it has been necessary to divide the wire at the loop, and to draw out the two portions separately; or to stretch the wire straighter by forcible traction at both ends previous to withdrawing it.

After the wire is withdrawn, the groin in the neighbourhood of the puncture must be compressed gently every day, to free the deeper parts of the wire track from pus. This must be done carefully, by pressing the groin deeply in a direction towards the puncture on each side, so as to get the pressure behind the wound as much as possible. The fistulous track may be afterwards washed out every day with zinc lotion. In some cases, where the granulations appeared weak and relaxed, I have used a solution of tannin, and also a lotion of the muriated tincture of iron, with great benefit. The tissues were rendered firmer and harder, and the tendency to their suppuration checked. I have thought that, in these cases, the succeeding induration was harder and more enduring. The upper puncture is usually the first to close, the lower acting for some time longer as a drain to carry off the discharges. This usually takes place in the third week, when a spica bandage and a pad of good size may be applied firmly, and the patient allowed to get up and walk about his room. As soon as the parts will bear it, a light truss of the proper shape may be worn without waiting for the complete healing of the lower wound.

The kind of truss I recommend, and have been lately in the habit of treating patients with after operation, is the horseshoe pad truss, of which a description will be found in the following pages. The spring of the truss must not be too strong; it should be strong enough to support but not to compress; and the pressure should be quite flat, and be spread over the neighbouring parts as well as the obliterated canal, so that the pad of the truss should not be too small. (See the section on *Trusses*.)

VIII.

VARIATIONS OF THE WIRE OPERATION.

In some of the earlier wire cases, various modifications in the application of the wire were tried. One was adopted in consequence of the length and pendulousness of the hernial sac.

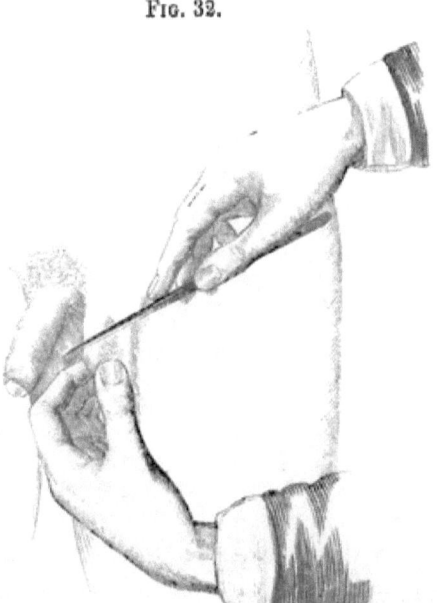

Fig. 32.

A great redundancy of the skin of the scrotum led me to remove altogether an elliptical portion with the knife, after pinching it up between the finger and thumb, in the manner shown in Fig. 32. Then, after separating the skin around the

120   ON INGUINAL HERNIA.

margins a little, the fascia was invaginated, and the needle first passed through the conjoined tendon as before described—the wire, with one end thinned off, hooked on above and drawn through into the scrotum. The needle, disengaged from the wire, was then passed across the sac in front of the cord at the scrotal incision, the lower end of the wire hooked on and drawn back with it across the back part of the sac. (Fig. 33.) The

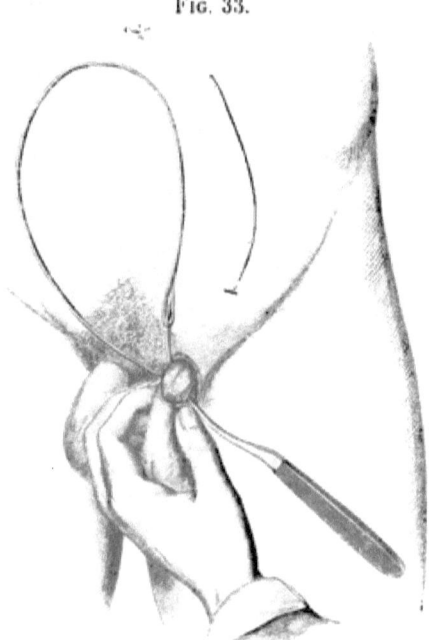

Fig. 33.

needle, carrying the thin end of the wire, was then passed in the usual place through the middle of Poupart's ligament, as seen in Fig. 34. The wire was then freed, and the needle withdrawn, leaving the two ends of the wire in the upper puncture; these were then crossed over each other and twisted round a wooden compress, as seen in Fig. 35, A, B.

Thus the loop of wire was sunk in the hernial canal, enclosing the conjoined tendon, the sac, and the two pillars of the superficial ring, invaginating the sac and scrotal fascia, and retaining them powerfully upwards against the internal opening of the hernia.

The above proceeding differs somewhat from the one before described. Carrying the needle fastened to the wire through Poupart's ligament is, perhaps, a little more difficult to accomplish. The use of the external compress is also retained.

Fig. 34.

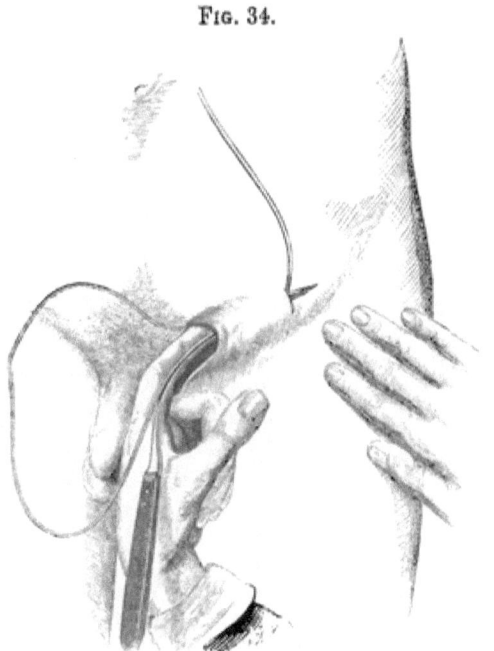

It is an operation to be preferred, perhaps, where it is desirable, from the length of the sac and of the hernial canal, to obtain greater upward traction of the former into the latter.

Another modification is represented in Fig. 36. It was rendered necessary by the great width and lax condition of the inguinal canal and external ring, amounting to a direct opening into the belly, and requiring an additional precaution for the closure of the superficial opening. After invagination of

ON INGUINAL HERNIA.

the scrotal fascia, the needle was passed, and the wire placed in the usual way through the two sides of the canal, as high up as possible, and the two ends of the wire drawn into the scrotal incision in the manner before described. The ends were then made to cross each other in the interior of the sac. This was accomplished by hooking the inner end of the wire (both ends of which must be drawn thin in this operation) on to the needle, pushing it across the sac from without inwards

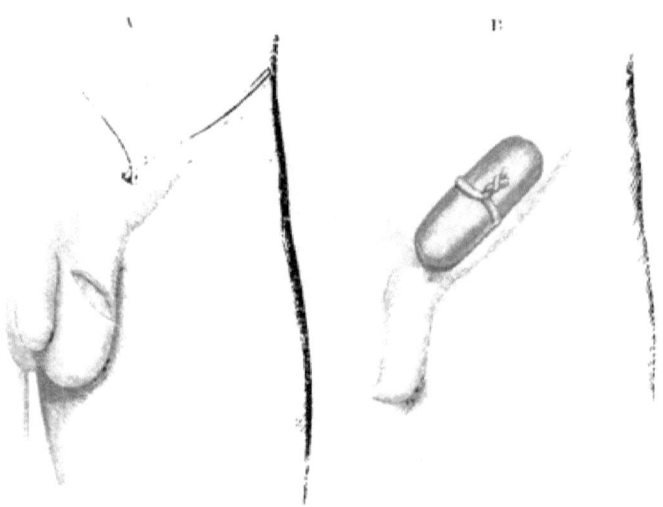

Fig. 35.

in the manner before described; and then, by freeing that end from the instrument and hooking on the outer end, which was then drawn through the sac in its turn by withdrawing the needle. The sac was then invaginated as much as possible by traction on the loop, the wires being made to hold by bending their ends forwards. The needle was then passed through the insertion of the internal pillar into the front of the pubis, and the nearest end of the wire is drawn back with the instrument. The same process was repeated at the insertion of Poupart's ligament, taking through it the remaining end of the wire on the inner side of the spermatic cord, which was pushed outwards during the manœuvre. The additional hold thus given to the wires upon the pillars of the superficial ring lessens the strain upon the punctures during the action of the recti, which, in large cases, has a tendency to split or fray the parallel fibres forming these structures. At the same

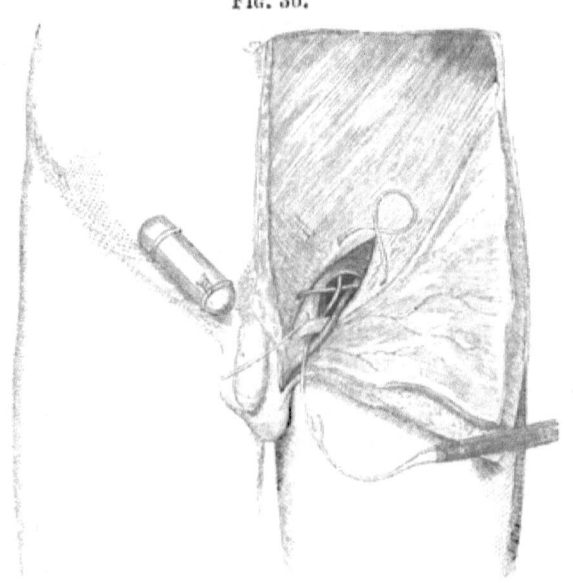

time, the hernial opening is closed at the lower part, close above the pubis, where I have found the greatest tendency for the rupture to escape after an operation, in large cases. In large and difficult cases, these advantages more than compensate for the additional time and trouble in operating. The pubic crest and spine are prominent guides to the operator in the latter part of the operation, and prevent any danger to the deep-seated parts from the point of the instrument. An elongated, cylindrical boxwood pad was then applied to the surface; the upper end was inserted into the loop of wire emerging from the upper puncture, and round the lower end were twisted the two free ends of the wire at the scrotal incision, as seen in Fig. 36, on the right groin. This method was employed in *Cases* 23, 37, and 59 (*Appendix*).

In Fig. 37 is seen a representation of the wires "in situ," applied in a large case of direct hernia. (*Case* 22.) It differs from the operation just described only in the two lower ends being applied to their respective pillars, on the opposite side, instead of the same side, of the spermatic cord. This is easily accomplished by pushing the cord outwards before applying the last end of the wire. By this means, the inner wire is made to traverse the hernial canal and sac twice, opposite to and below the deep opening, and in front of the spermatic cord, which is also crossed twice by the wire.

By these two operations, intended for very large cases, we obtain additional holds upon Poupart's ligament, at its insertion into the pubic spine, and upon the internal pillar of the superficial ring, and tendinous origin of the rectus muscle. The tendency of these herniæ to escape close above the pubic crest is thus provided against. In very large and lax cases, the additional employment of the compress is desirable; though, if preferred, the loop and ends of the wire may be twisted down into their respective apertures of egress, and then united over a rolled pad of lint, as usually employed.

Mature observation of the effects of the external compress has led me, in ordinary cases, to prefer the method of twisting up the wire into the punctures.

The effect of the compress is chiefly exerted upon the superficial parts, consolidation or effusion in which has no beneficial effect in imparting a resisting power to the inguinal canal. At the same time, the pain and uneasiness to the patient are considerably increased by it. In most cases, it has a tendency also to prevent the free escape of the discharges, and usually requires an early removal for this reason. In other cases, it becomes speedily so loose by the ulceration of the wire track as to be of no use after the first week. Its effects will be seen to be chiefly felt by the tissues which are superficial to the wire or ligature, and are included between them and the compress;

FIG. 37.

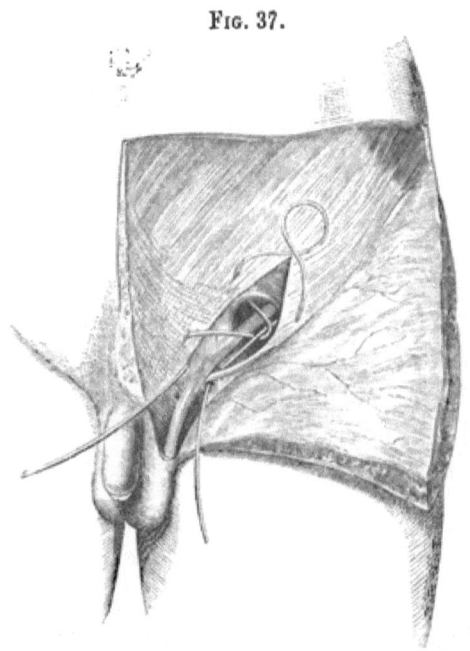

while it tends to prevent an effective pressure by the wire upon the deeper parts of the canal, in which adhesion is the most essential to success.

The twisted wires, on the other hand, favour the escape of the discharge along them, and serve to transmit the pressure of the bandage to the posterior or deep wall of the hernial canal, along which the rupture is apt to make its way and recur.

The facility with which the wires are usually untwisted and

removed, on account of the enlargement of the track along which they lie by ulceration, removes a theoretical objection which I at first felt against this mode of fixing them, on the score of the difficulty of getting them out of the wound after the parts had become consolidated.

At the same time it must be said, that the action of the compress is beneficial where there is great laxity about the superficial ring, upon which part its effects may be considered as more direct and certain than upon the deeper structures.

Before the acquisition of experience had proved the feasibility and advantages of twisting up the wire, one case was operated on by the following method.

The usual preliminary scrotal incision and invagination of the fascia being effected, and the wire applied to the conjoined tendon and Poupart's ligament in the manner described, the upper loop was drawn entirely into the groin puncture, so as to lie under the integuments close upon the aponeurosis. The ends of the wire in the scrotal incision were then held together by a sort of clamp, made of a thin elongated piece of steel, with a loop or ring at the upper end and a cross piece at the lower to fasten the wire upon. In fixing it, the ends of the wire were passed through the ring of the clamp, which was then slid over them up into the wound, and made to invaginate the fascia as high up into the canal as possible; the ends of the wire being finally bent round the cross piece at the bottom, and so fastened.

In the case on which this plan was tried (*Case* 25), the upper puncture healed completely over the wire by the adhesive process. The discharge which resulted, not being able to escape freely enough by the lower incision, accumulated in a small abscess in the groin over the loop end of the wire, and required an incision for its escape. The pressure of the steel clamp in the canal resulted in the sloughing of the sac in the scrotum. The ultimate result of the action thus set up around the cord was considerable atrophy of the testicle.

These serious drawbacks led me to discard the use of the clamp in succeeding cases, and to resort to twisting of the wire above and below, and to the maintenance of the upper as well as the lower opening by the twisted ends of the wire. This provision for the free escape of the discharge has been so effective, that I have had, since the time I adopted it, no instance of retention of matter or the formation of an abscess in the groin or deeper parts, and in the scrotum but rarely; nor of any atrophy of the testicle, except in one instance from other causes.

Lastly, in a very large case of congenital rupture (*Case* 48), in which the hernial apertures were directly opposed to each other, and the previous use of the rectangular pins had obliterated the scrotal portion of the sac, and reduced the opening to a chink at the edge of the rectus muscle, an operation was performed, the arrangement of the wire in which is given in Fig. 38. In this case, no incision was made in the scrotum, the coverings of which were so thin as to permit distinct tactile perception by simple invagination of the skin.

A common stout suture needle, large enough to carry a wire somewhat thinner than that ordinarily used, was employed. The little finger of the operator being firmly invaginated as high up as the aperture would allow, the skin of the groin was drawn well upwards by an assistant. The needle was then passed from the surface of the groin through the skin and outer pillar, on a level with the upper part of the hernial aperture. The point of the needle, being felt to impinge upon the apex of the invaginating finger, was then carried across the canal, and made to transfix the internal pillar, and the covering of the rectus muscle placed behind it. It was then pushed through the skin over the internal pillar. A short length of the wire being thus drawn through, was cut off and left in the puncture. The skin of the groin was then drawn downward, the invaginating finger receding a little at the same time, and placed firmly upon the spermatic cord behind the extremity of Poupart's ligament. The needle being passed through the same external aperture as before, was again made to transfix Poupart's ligament upon the end of the finger, and carried across

FIG. 38.

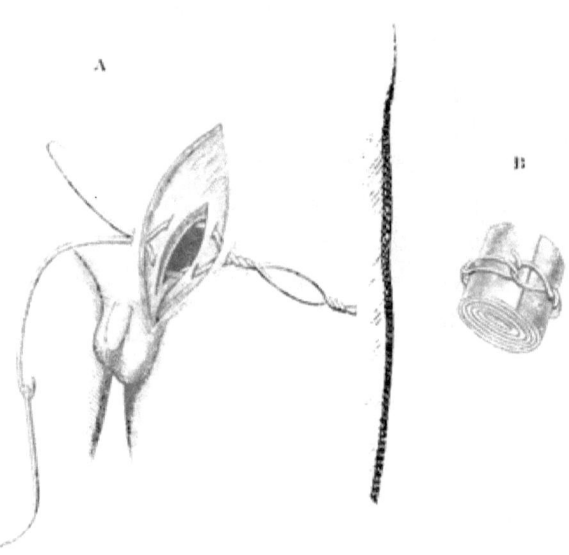

In this figure an opening is represented in the skin and aponeurosis of the groin, in order to show the arrangement of the wires in the canal.

the cord to the edge of the rectus at its attachment into the pubis. Through this and the overlying internal pillar its point was next thrust, and carried through the skin-aperture before made in this situation. Another length of the wire was then cut off, and the skin of the scrotum suffered to drop out as much as it would. The extremities of the two wires at each puncture were then twisted firmly up, with the effect of completely closing the aperture in its whole length. At one end a loop was made by twisting, as seen in figure (A), and at the other end of the wires two hooks were bent. The hooks were then fastened to the loop over a pad of lint (B). Instead of the pad of lint, a cylindrical boxwood or glass pad may be used in preference, in large cases, as in Fig. 36.

IX.

OPERATION FOR CONGENITAL HERNIA IN CHILDREN BY THE USE OF RECTANGULAR PINS.

In congenital hernia, and the smaller kinds of rupture in children and young boys, the hernial canal is usually narrow; the internal opening often contracted; and the sides elongated so as to retain much of their valve-like action, and so flexible and yielding as to admit of easy approximation in their entire length.

Coupled to these advantages, the scrotal fasciæ and coverings are thin, and the fundus of the sac often small, so that the tissues between the skin and the sac do not constitute any great bulk, and are consequently of little use as a connecting medium, when transplanted into the hernial canal. The flexibility of the canal itself, also, is such as to need no interposed tissue to fill up the hernial openings; while the thinness and delicacy of the scrotal tissues permit of a distinct perception of the anatomical relations of the deeper inguinal structures by simple invagination of the skin, and without a preliminary incision.

To draw together and cause to unite the sides of the canal under such circumstances, I have used, in thirteen of the cases given in the Appendix, the *rectangular pins* which I had been for some time in the habit of using for the cure of varicose veins and varicocele, by double subcutaneous acupressure. For the cure of rupture, however, these pins are made a little larger and stronger than for the purpose just named, and both are curved considerably more near the point.

The pins are represented in action in Fig. 39. They are of

OPERATION BY PINS.

various sizes, from three to five inches long, and of proportionate strength; hardened at the point, but made in the shaft soft, so as to bend rather than break off short under tension.

They are spear-pointed, flattened on the concavity, with slightly cutting edges, and curved boldly, like the ordinary suture needle, near the point. The shaft is rounded, and at about three or four inches from the point it is bent at right angles, the angle being twisted into a loop large enough to receive the point of the fellow needle. The projecting end is about an inch long, and terminates in another loop, for the purpose of securing it in position by the dressings. When about to be applied, the pins are held by the middle of the shaft between the thumb and middle finger, the last phalanx of the forefinger being placed upon the bent extremity, with the terminal loop towards its point. (See Fig. 40.)

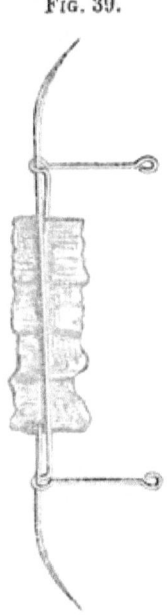

Fig. 39.

They are applied separately, and in opposite directions, through the structures intended to become adherent, on each side of those intended for obliteration; and are pinned through and through as in fixing an ordinary toilet pin. Each point is then passed through the loop at the angle of the fellow pin; and the bent ends afterwards rotated in opposite directions upon the axis of each other's shafts, so as to twist up and compress the included structures, somewhat like the twist given to the neck of a mail-bag.

The Operation.—The method of applying them to a case of rupture, as I have most frequently practised it, is as follows:—the hernia being reduced completely and carefully, and the child laid on its back, with the knees bent up; the scrotum is invaginated into the canal, and the position of the cord, the conjoined tendon, and Poupart's ligament, carefully recognised and noted. In small cases, this can be most conveniently done by using the little finger. This is carried up into the canal as far as possible, until the border of the internal oblique muscle can be felt as a thick covering in

FIG. 40.

front of the point of the finger. The conjoined tendon may then be made, by traction, more salient and evident at its inner border.

The skin of the groin being drawn *directly* inwards by an assistant, a pin, with its concavity directed downwards, is passed through it, and all the intervening tissues, straight upon the inner side of the nail of the invaginating finger, till it can be felt to touch it. (Fig. 40.) It is then slid downwards

OPERATION BY PINS.

along the side of the finger, which is at the same time withdrawn along with it, protecting it on the outer side until the point of the pin can be felt to touch the upper border of the pubis. Skirting this bone, it is then carried down into the scrotum, and brought out through the skin over the fundus of the hernial sac, upon which the invaginating finger has been pressed. In this manœuvre the points of the finger and pin must be kept fairly " pari passu ;" and the rupture kept

Fig. 41.

out of the sac, by the pressure of the little finger of the hand carrying the pin, over the internal ring. This position will, at the same time, give steadiness to the hand.

The second pin is then taken, and its point, with the concavity forwards, passed through the skin into the same scrotal puncture at which the former emerged. The invaginating finger is then placed below it, and the scrotum again pushed up before it into the canal, carrying the point of the pin upon it till it touches the posterior surface of the outer pillar of the superficial ring. Through this it is then passed obliquely upwards and outwards, so as to raise the skin of the groin directly below the situation of the deep ring. The skin is then drawn outwards by an assistant as much as possible, when it will be found that, by turning the concavity of the pin inwards, its point can be brought out through, or sufficiently close to, the puncture made by entering the first pin. (See Fig. 41.)

Fig. 42.

The pins are then to be locked into each other's loops, a proceeding which will be much facilitated by first cutting off the sharp end of one of them with pliers, so as to make it shorter than the other. The point of the other pin is then to be removed by the pliers, and the bent ends twisted round once. This may be accomplished best by simply turning over the end of the upper pin towards the thigh. The punctures and skin

OPERATION BY PINS. 135

are then carefully protected, by lint or wash-leather and adhesive plaster, from the effect of pressure by the loops and ends of the pins, and the extreme loops of the pins are tied together by string, or a strip of lint or sticking-plaster, so as to prevent slipping. (See Fig. 42.) The whole is then retained by a pledget of lint and spica bandage.

Fig. 43.

In this operation the conjoined tendon and the internal pillar are transfixed by the pin first applied, and the outer pillar is transfixed and included by the second. The sac is transfixed by both the pins, which lie for some distance in its interior. The whole of these parts are twisted together by the revolution of the one pin on the axis of the other, so that the posterior wall being drawn forward, and the anterior drawn backward, the hernial canal is firmly closed; tightly embracing the cord, which lies behind and between the pins, but without being included by them.

In one case operated on (*Case* 31), the pins were placed *transversely* across the canal,—one deep, so as to lie close upon and across the cord (as shown in Fig. 43); and its fellow,—superficial to the external oblique aponeurosis, and under the integuments only. (Fig. 44.) When only a couple of pins are applied in this way, however, the power of twisting together the sides of the canal in its whole length is not obtained, and it is closed only at one point, constituting a less resistant barrier against future protrusion. A want of complete success,

Fig. 44.

which followed this plan in the case referred to, led me to follow it only as an additional precaution to the operation first described, in instances of large herniæ requiring it.

In the instances last mentioned (*Cases* 47 and 48) the ruptures were congenital, of great size, and the openings so large as to admit easily of three fingers directly into the abdominal cavity, close above the pubis. No truss had been obtained which could keep up the bowel at all, and the little patients (brothers, of the ages of four and a half, and seven years respectively) were in a deplorable condition in regard to their future prospects of maintaining themselves by labour.

Under these circumstances, I combined the two operations by fixing a couple of pins of the larger size, in the manner first described; and afterwards passing another couple of the smaller pins transversely across the sac and pillars, close above the pubis. The deeper of these two was passed close to the bone, transfixing both pillars of the ring at their insertion upon it; the more superficial one passing under the integuments only, through the same skin punctures. Enclosed between them were the larger pins first applied along the sides of the canal.

The result of these two cases has been, so far, very satisfactory, requiring only a little additional interference afterwards to close up (in one case completely) a hernial aperture so large as to permit the finger to traverse almost the whole of the lower part of the peritoneal cavity. Not a single unpleasant symptom occurred in either of the cases afterwards.

The pain experienced by the little patients after the operation by pins was uniformly so slight that no sedatives were employed. The tongue was barely, or not at all, whitened; the appetite undiminished, except for one day or so after the operation (and this might be properly attributed to the chloroform); and the action of the bowels unaffected. In no case was there any suppuration worthy of note; in the smaller cases, none at all. A drop or two of serum oozing from the punctures and crusting the dressings, was usually the only discharge.

The dressings were changed only once or twice in the course of the treatment, and then more to ascertain the progress of the consolidating action and the state of the punctures than from any real necessity. The little patients suffered so little that they were with difficulty kept still enough, or in the recumbent posture. The pins were usually withdrawn, without difficulty and with little pain, on the tenth day. In one or two cases they were withdrawn about the fourth or fifth; and, in another, on the second day, in consequence of the amount of consolidation and thickening set up in that short time. In one instance they were withdrawn early, because of the occurrence of some symptoms which were thought to betoken the access of scarlatina. (*Case* 39.) All these cases were perfectly successful. The case first referred to (*Case* 31) was not successful, chiefly, I believe, in consequence of a bad state of health from bronchitis, teething, and general tubercular diathesis. A violent cough which supervened gave rise to a partial return of protrusion; but this, by the pressure of a proper truss, was much more under control than before the operation. The same child was operated on on the opposite side (the rupture being a double one) with uninterrupted success. In another case (*Case* 34) I thought there had occurred an increase in the impulse at the site of the operation the last time I saw it, but this was easily controlled by the truss.

A hard cord, and sometimes a great amount of thickening, could be felt in the canal after the withdrawal of the pins. The fundus of the sac could be felt obliterated and hardened into a solid mass of effusion within the scrotum, while the superficial ring could scarcely be distinguished. In some of these cases the testicle was somewhat enlarged for a while, but gradually recovered its normal condition.

The foregoing operation must be viewed rather as an adjunct to exclude the bowel, and so to accelerate and render certain the treatment by the pressure of a truss, with a pad constructed on the proper principle for closing up the sides of the canal. Besides obliterating the interior of the original sac, the action of the pins on the canal gives an additional tendency to that closure around the cord which should normally succeed the descent of the testicle. The pressure of a truss-pad, of the proper shape, will doubtless often accomplish the same result, but by a much longer, more tedious, and more uncertain process; while the use of an improper truss-pad, instead of accelerating a cure, will increase the size of the hernial openings and make the state of things worse, as is commonly met with in practice.

An operation so little painful, so simple, and so totally free from danger when skilfully conducted, does not seem to be open to any of the objections commonly urged against operative interference in cases of congenital hernia in children; while it renders the certainty of a cure so much greater, and its accomplishment withal so speedy, as to constitute strong recommendations to its adoption; especially in cases where the prolonged friction of a truss strong enough to retain the rupture, is followed by frequent sores, gallings, and abscesses in the delicate groins of infants.

X.

THE MODUS OPERANDI OF THE FOREGOING OPERATIONS.

The results aimed at in these operations may be briefly recapitulated in the order of the importance attached to them respectively.

First.—The posterior and superior boundaries of the dilated tendinous canal are drawn forwards, downwards, and outwards, and become permanently united by adhesion to the anterior and inferior walls; or, in other words, the conjoined tendon, the internal pillar of the superficial ring, and, in some cases, the edge of the rectus tendon, are firmly united to Poupart's ligament and the external pillar of the ring. This adhesion follows upon the close compression and slow ulceration induced by the action of the wire, thread, or pins. In Fig. 45 it will seen, that the effect of this adhesion is to restore the retaining power of the valve of the posterior wall (*v*), by approximating it to the anterior (*a*), in cases of oblique hernia. The contraction of the cicatrix, subsequent to the operation, promotes the closeness of the apposition and increases the future security against protrusion. The spermatic cord is closely embraced by the contracting tissues as it lies in the groove above Poupart's ligament, the latter protecting it from undue pressure.

Fig. 45.

In cases of direct hernia, this restoration of the overlapping valve is necessarily imperfect, since there is a deficiency here in

MODUS OPERANDI OF THE OPERATIONS. 141

the limb of the valve itself. Here the ligature acts mainly as a circular compressing force upon the opening, uniting Poupart's ligament and the outer pillar, to the edge of the rectus tendon, the inner portion of the conjoined tendon, and the internal pillar. For greater security in ruptures of this kind, an additional point of union of the pillars of the ring close to the pubis becomes necessary, especially where the opening is large; and the employment of some of the modifications of the wire operation given in the preceding pages will give rise to a greater extent of adhesion and consolidating action.

Secondly.—The consolidation and contraction of the tendinous structures traversed by the ligature or pins under the healing process which succeeds the withdrawal of the latter, render them more adherent to each other, and more resistant to pressure from within. This effect will be better understood by reference to the adjoining figures, showing, one (Fig. 46) a vertical, and the other (Fig. 47), a horizontal section of the parts opposite the cicatrices. The contraction of the circular cord of effused fibrin will draw the skin at *a*, towards the deeper parts and peritoneum at *v*; enclosing and compressing the canal,—thus supplementing the retaining power of the "*arciform*" and "*intercolumnar*" fibres. Thus Poupart's ligament becomes a "point d'appui" for the compressing forces, and a direct opposition is afforded to the yielding of the superior and posterior walls. The effect is also to resist that dilatation of the canal, before pointed out as produced in hernial cases by the action of the recti muscles, by transmitting their traction force along the cicatrix to Poupart's ligament, and aponeurosis

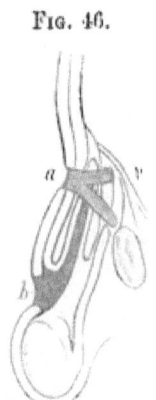

Fig. 46.

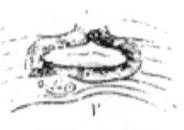

Fig. 47.

of the external oblique. The mobility of the different abdominal layers upon each other, which has been before adduced as a cause of the more easy formation of a rupture, is also limited and restrained by these adhesions.

Thirdly.—The transit of the ligature or pin across the sac, and along its interior surfaces, sets up adhesive action there, which results in its entire obliteration and adhesion to the spermatic cord, as in the treatment of hydrocele by seton. Combined with the compression exercised upon the neck of the sac, which throws the serous membrane there into a multitude of folds, and with the action set up in the surrounding tendinous structures, the result is the formation of one mass of adhesion, which totally precludes the formation of another rupture in the same sac.

In the small cases before referred to, in which the sac is not implicated at all, but pushed up into the deep ring, this obliteration of the sac is not at once effected by the operation, but is accomplished by the slower process of contraction through the elasticity of its serous membrane and the absence of distending force from its interior.

In older and larger cases, where the sac has formed adhesions in the scrotum, the obliteration of its cavity becomes a matter of more importance, as it then forms a support to the adhesions at its neck in resisting a return of the rupture.

In one or two of the cases given at the end of the volume, it will be seen that a liquid effusion into the posterior fold of the fundus of the sac followed the obliteration of its neck by the suture. That part of the sac which became the seat of this effusion was thus in a condition similar to a hydrocele. The most evident course was to treat it on the same principle.

In one, the liquid serum was drawn off by puncture with a grooved needle, with the best results. (*Case* 30.) In another, the effusion spontaneously disappeared, the case making a good cure. (*Case* 1.) In a third, the effusion continued, combined with some varicocele, up to the time I lost sight of the case. (*Case* 5.) At that time a communication evidently still existed between the sac and the peritoneum, since the fluid could be pressed very slowly upwards, and the greater part of the tumour could be made to disappear. The opening, however, must have been small, the sensation given to the compressing fingers being exactly like that communicated by an india-rubber bottle from which the contents are escaping by a small opening. Much induration was present about the neck of the sac and inguinal canal.

Since such a condition might result in a re-dilatation of the hernial passage, I should not hesitate, at the present, as I did at that time, to puncture the sac and let out the fluid.

A short time ago I had sent to me a most interesting case of rupture, the condition of which had an important bearing upon the present subject. The rupture was large and scrotal. A hard mass, evidently omentum, could be felt in the inguinal canal, immovable and adherent to the neck of the sac. The fundus was distended by a translucent serous effusion, which had all the characters of a hydrocele, except that the testicle was entirely distinct, though considerably enlarged. It could be freely moved without dragging the swelling along with it.

The history stated that the rupture was of one year's standing, and had been scrotal for some time, but was easily and completely reducible until a week previous to his application. At this time, in consequence of wearing a powerful but ill-fitting truss, much pain and great increase of swelling occurred in the scrotum, and it was found impossible to return the rupture as before. The bowels, however, continued regularly to act. A puncture was made in the swelling, and about three ounces of serous fluid, highly albuminous, escaped. The omentum blocking up the neck of the sac could then be distinctly felt just above the superficial ring. The patient was much relieved by the operation, and I advised him to have nothing further done to it at present, since there was a strong probability that a cure would be the result by adhesion of the omentum to the sac.

In three cases (Nos. 8, 25, and 52) the sac suppurated and sloughed below the neck, the ultimate result being in one case highly satisfactory, and in the other some atrophy of the testis. This result, however, has been quite exceptional; and only in the last case has it followed the many operations by the most approved methods of using wire or pins. In the case alluded to, sloughing was endemic in the hospital at the time.

Fourthly.—When the scrotal fascia is invaginated into the hernial canal, the adhesion of its opposing raw surfaces to each other, uniting with the general mass of consolidation in the track of the ligature, and to the surface of the spermatic cord, forms an additional provision against a return of the rupture. (See Fig. 16.) In very large ruptures this fascia, being commonly very strong and hypertrophied, contributes beneficially to fill up the large internal opening at the neck of the sac, and serves for a connecting bond of union to the widely separated boundaries of the opening; affording, at the same time, a convenient bed or point of resistance to tighten the ligatures upon. The fascia of the scrotum is extremely vascular, and apt to take on a consolidating action, in the fibrin which is abundantly effused in the presence of a foreign body. It is at the same time very tough and resisting when condensed by inflammatory changes. Thus, in large ruptures, it assists in maintaining the closure of the canal.

The new adhesions which are formed by the edges of the scrotal incision to the margins of the superficial ring, also contribute powerfully to support in its new position the invaginated scrotal fascia against pressure from above.

Fifthly.—The union of the pillars of the superficial ring to each other, either directly, as after the use of the pins, or through the medium of the invaginated fascia, as in the other operations, has also a powerful influence in supporting the adhesions formed higher in the canal. In direct hernia this result becomes especially valuable, increasing the limited area of adhesion around the neck of the sac. It will be remarked that this union, although one of the most original of the features of the operations devised by the author, cannot be effective in itself and alone, in permanently closing the hernial canal, since it would be simply the conversion of the hernia into a bubonocele, while the limited area of adhesion would give way anew upon any great force acting from within.*

The closure of the superficial ring can be distinguished in most of the cases after operation. A firm fibrous band passing from one to the other, and adherent firmly to the cord as it emerges, resists the passage of the surgeon's finger into the canal. Passing down from it to the cicatrix in the scrotum can usually be felt also a firm cylindrical fibrous mass, the results of the invagination of the scrotal fascia and its firm union to the sides of the ring. Such a fibrous band can be discerned also in the cases treated by the pins. In some cases it is formed by the remains of the fundus of the sac remaining in the scrotum, shrivelled, and contracted into a hard, firm cord. These

---

* The operation proposed by Professor Chisholm, of Charleston, America, of applying subcutaneously small wire sutures to the pillars of the superficial ring, and leaving them permanently under the skin, is not likely to result in permanent benefit, for the reason given in the text. Under the pressure of the bubonocele which is left in the canal, the small wire thread will cut like a knife the tense fibres of the pillars, and the rupture will again protrude, carrying the sutures on its anterior wall. It should, in justice, be here mentioned that the above-named gentleman, when in this country some years ago, and before his cases were operated on, was present at a demonstration by the author of his operation on the dead body, at King's College; when the possibility of such an application of the sutures was discussed, and the reasons against its adoption given in the text were stated.

conditions have been detected in many cases years after the operation. Any invagination of the skin itself between the pillars of the external ring in the operation will, of course, prevent this union of the sides of the opening, and must be carefully guarded against.

It may be found, by some, difficult to understand how such union can be produced, without denudation by the knife of the surfaces of the tendinous and other structures. The action of the suture and pins upon the tissues which they traverse, and which lie in their grasp, is that of slow division by ulceration and absorption. The effect of this is to produce a denuded surface in a subcutaneous position, and with but little, if any, exposure to the atmosphere. The adhesive process, or that of granulation, consequently follows directly and closely upon the track of division of the tissues, and is uninterrupted by the accidental complications which commonly follow an exposed wound. An additional advantage is derived from the fact that the parts are held and retained in the position which it is desired they should maintain, by the suture or pins, until adhesion and granulation have accomplished a more permanent retention and occlusion of the abnormal openings.

XI.

CAUSES OF FAILURE AND DANGER IN THE OPERATION.

THE main cause of the repeated failures in the ancient and modern operations to close up the hernial passages, lay in the want of recognition of the necessity of including the posterior wall of the canal in the parts operated on. The anterior parts of the hernial coverings only being influenced, the rupture generally descended *behind* the site of the operation.

This difficulty may be also experienced by a surgeon, unaccustomed to manipulate the parts, in the performance of the foregoing operations, hesitating at the dangers of wounding the epigastric artery, the peritoneum, the spermatic cord, or the bowel, when fixing the sutures.

In some cases, the conjoined tendon is not well marked, the attachment of the fascia transversalis obscuring its outer border. If, in these cases, however, the finger be well pushed up behind the lower border of the internal oblique, and then flexed forcibly towards the surface, the tendon sought after may generally be distinguished as a ridge or border placed on the stretch. The point of the needle should then be passed inwards, so as fairly to take it up. It may be thought that the epigastric vessels are endangered at this part of the proceeding. An inspection of the parts in a hernial subject will show, however, that the finger pushed upward between the internal oblique muscle and fascia transversalis, opposite the deep ring, will press back the fascia covering these vessels and the peritoneum, forming for them a safe protection; so that they will be out of the reach of the needle placed properly on the inner side of the invaginating finger. The exercise of a little delicacy and experienced tact is certainly required in taking up the requisite amount of the tendon and no more. The edge of the rectus tendon, which may be distinctly recognised, will indicate the depth as well as the distance inwards, to which the needle may with safety be passed. In direct cases, this portion of the rectus may be partially included in the ligature.

The shape of the point of the needle, as before described, is such as to split rather than cut across the tissues, which may result if there be any cutting edge on the shoulders. A neglect of this precaution in one operation I witnessed, led to a complete division of Poupart's ligament, as if with a tenotomy knife. The same result may follow the use of thread or wire too thin; or a too great tightening of the sutures. In one of my own cases, such a cause led to the enlargement of the needle puncture so great, as to permit of a small protrusion after the operation. This, however, was diminished by the pressure of a flattened truss pad. The blade of the instrument should be well and sharply curved towards the handle, so that when deeply inserted into the parts, the point has a tendency to come short to the surface. When the point is fairly implanted behind the tissues intended to be included, the curve of the blade should be placed across the invaginating finger, which may then be used as a fulcrum upon which to turn forward the point, by depressing the handle. At this time the invaginating finger may be itself brought forwards, carrying the needle upon it. The concavity of the latter should be kept carefully and firmly turned forwards and a little upwards during the performance of these manœuvres.

A similar manœuvre performed behind Poupart's ligament will guard the iliac vessels and cord, by lifting up the aponeurosis forming the external pillar away from them, during the passage of the needle. A firm pressure by the thumb of the invaginating hand upon the femoral vessels at the same moment of time (as seen in Fig. 26), will render the precautions certain.

It may be pointed out here, in reference to the danger menacing the vessels and peritoneum, that in none of the cases given in the Appendix has there been any hæmorrhage. In one case there was a slight venous oozing, evidently proceeding from one of the spermatic veins, which were varicose. In the two cases in which alone some degree of peritonitis was suspected, the symptoms did not appear until the lapse of a week, and were caused, as I believe, by the burrowing of matter towards the peritoneum.

In some operations for the radical cure of rupture, I have heard of the bowel having been pierced, and even tied up in the suture; the result being, as may be anticipated, death in a very short time after the operation. Unless the operator be rash and over-confident, or the patient struggles so much as to bring the bowel forcibly down at a critical moment unprepared for by the operator, such a disastrous result ought not to occur. The complete return of the rupture, and the recumbent position of the patient, with the pelvis raised, generally suffice to prevent trouble from the presence of the bowel in the sac. I place the most reliance, however, upon the firm pressure of the invaginating finger in the neck of the sac blocking out the bowel, and giving a clear perception of the tissues about to be perforated. If due care be taken to hook the point of the finger firmly forward and a little towards the side of the canal about to be punctured, so as to feel the parts fairly stretched over it; and, at the same time, to be sure that the point of the instrument is not permitted to pass beyond that of the finger towards the inner opening of the sac, the greatest confidence may be felt that no bowel can be included.

When the patient struggles much under the chloroform, it

is better to wait until its full effect is produced, if it can safely be done. If not, an assistant should command attentively the internal ring by pressure with his fingers, whenever the steps of the operation will permit him to do so. If coughing occur, the needle should be kept out of the canal, or quickly but carefully withdrawn if already placed there, especially if there be a tendency of the bowel to pass down beside the finger in a wide canal. If the point of the needle, however, be at the time fairly engaged in the walls of the canal, the occurrence of the cough can do no damage, and the operation may be calmly proceeded with. In some very obstreperous cases in children, I have had recourse to *inversion* of the patient, by placing the pelvis higher than the head, as a precautionary measure, which had all the effect intended. I found this plan of placing the patient topsy-turvy very effective also in the succeeding dressings, when it was necessary, during the change of the compressing agents, to prevent a forcible disruption of the newly-formed adhesions, by a violent pressure of the intestines from within during the struggles of the patient. (*Cases* 46 and 47, *Appendix*.)

If the surgeon has sufficient tactile perception of the structures in the canal, a proper amount of caution and discretion, a steady hand, and that precision and confidence in the use of the instruments, which can only be obtained by experience, there is, I conceive, but little danger of mischief being done. I say this after performing the operation sixty-six times on the living, and a great many more on the dead subject.

At the same time, the possibility of passing the needle into the peritoneal cavity, of including the epigastric or wounding the iliac vessels, of transfixing the body of the rectus muscle, and even of transfixing and including the bowel, must not be forgotten by the operator.

The fear of passing the needle too deeply may cause the

conjoined tendon to be imperfectly secured. Many failures have resulted from this cause. With a view to lessen the chances of such failure, I have employed and recommended in the wire operation as well as in that by pins, the application of external pressure by a pad and spica bandage upon the wire loop or pins. When such pressure is made upon the stiff and unyielding loop formed by the twisted wire, bent over a pad of lint, it is communicated to and mainly exerted upon the deeper parts behind and in contact with the wire. Adhesive action is thus set up in that part of the hernial sac and canal which may have escaped the grasp of the wire, but which still remains in contact with, and is crossed by, its deepest part. By this direct pressure is secured adhesion to the spermatic cord, and complete obliteration across the hernial tube at the deepest and highest part traversed by the wire. (See Fig. 31, B and C.)

This direct pressure is obtained also, in a great degree, after the application of the pins. It is much more effective on that part of the canal which it is so important to secure, than any pressure which can be applied by a hard compress on the surface of the groin, the effect of which is experienced by the more superficial structures which intervene between the ligature and the compress, while scarcely any pressure is exerted upon the deeper parts. In some cases, however, in which much laxity is present about the superficial ring, a compress may be used with advantage as an addition to the wire. The advantages of both may be thus combined.

In some cases of oblique hernia which have returned after operation, I have observed that the first reappearance of the protrusion took place close to Poupart's ligament. The rupture afterwards crept along the course of the spermatic cord to the superficial ring, which became thenceforth slowly dilated. This result I attributed, in some cases, to a mal-position of the outer end of the suture, *i.e.*, its not being placed sufficiently far from the border of the outer pillar, or sufficiently close to Poupart's ligament. The width of the external pillar varies much in hernial subjects. In the smaller cases, the pillar is usually felt broad and distinctly triangular in form, preserving more or less its normal shape. In such cases the needle must be carried far outward behind the tendon. In other cases, the outer pillar is very thin and narrow, and can scarcely be distinguished from Poupart's ligament. In these the needle must be made to split the fibres of the latter structure itself, and the precautions before detailed to avoid wounding the femoral vessels must be carefully practised. In large cases, the outer pillars may be so lax and loose as to be distinguished with difficulty from the proper hernial coverings, and care must be taken to include the whole of the relaxed fibres in the suture.

Another source of failure consists in not placing the sutures opposite the exact centre of the openings which are sought to be closed. If too low, the hernial opening may be imperfectly closed above, or the contrary fault may permit the rupture to escape below. The latter is, perhaps, the more common result. The size of the opening will indicate to the surgeon the propriety of applying one or more points of suture on each side.

In the operations by wire, a more extensive approximation of the sides and openings of the canal is attempted by placing the wires obliquely in reference to the axis of the canal. (See Fig. 31, u.) If the wire be placed too low, a good deal of the effect of this obliquity is lost, and a bulging condition of the groin at the site of the internal ring remains after the cure has been effected. In these cases an insufficient action has ensued opposite to the internal ring, and its sides have not become adherent to the surface of the aponeurosis which covers it in front. If, on the other hand, the wire be placed too high, when twisted up tight it leaves the lower part of the canal comparatively unclosed, and the rupture may make its way behind the adhesions set up by the operation. This may be considerably, if not altogether, remedied in the latter stage of the operation on passing the wire across the hernial sac at the scrotal puncture, when a portion of the outer or inner pillar, at their insertion, may be included with the sac and its coverings.

It will be seen that in effecting this part of the operation, the canal being already closed more or less completely by the wire already placed on each side, the intestines are effectually prevented from passing into the sac, and getting into the way of the needle. This may be still further secured by holding the ends and loop of wire firmly together during the passage of the needle across the sac.

The most likely mischance to occur here, is the puncture or enclosure of the whole or some part of the spermatic cord in the grip of the suture. To a surgeon accustomed to operate for varicocele, this danger will seem but small, the sac of a hernia not being so close to the cord as the veins are to the spermatic duct. The cord can always be readily felt, and slipped back between the finger and thumb; and before the wire is tightened the cord can be distinguished as moveable freely behind it, and thus the operator is satisfied as to its safe condition.

In large cases, where it is necessary to pass the wire a second time through the pillars of the ring, the cord must be carefully distinguished, and pushed away from the side about to be punctured, before each application of the needle. The wire may be thus made closely to *surround* the cord without enclosing it. (See Fig. 28, B.)

If ligature thread is used, the success of the operation almost entirely depends upon the precision with which it is made to hug the spermatic cord, since the effect of the external compress is to make traction towards the surface of the groin, and *away from* the cord and posterior wall.

Failure may result in large and long-standing cases, from neglect in passing the wire fairly across the interior of the sac, the result being that the cavity of the sac in the scrotum is not obliterated, although the internal communication with the abdomen may be more or less completely closed. There is thus a want of support from below to the adhesions which have formed at the neck of the sac; or the fundus, as before mentioned, may become the seat of liquid effusion, which may increase so much as to redilate the canal from below.

Ruptures of the direct kind are more difficult to cure, and more apt to return after operation than the oblique variety, especially if the internal opening be large. In these difficult cases, the additional strain upon the adhesions binding the conjoined tendon to Poupart's ligament during the action of the recti muscles, after the operation, should be met by an additional hold by the wires upon the lower part of the pillars of the superficial ring; so as to ensure a more extensive permanent adhesion, as well as to lessen the immediate strain upon the upper point of suture, which might result in a fraying or splitting of the edges of the opening. (Figs. 36 and 37.)

Such herniæ usually present short necks and wide lax openings, and are generally found intractable to treatment by trusses. Cases 2, 10, 21, 28, 37, 38, 40, 52, and 53 in the *Appendix* are specimens of this variety, fortunately not the most common. In three, I failed in accomplishing, ultimately, a satisfactory cure, though the size of the hernia was much reduced, and the patient rendered more comfortable by the truss being made available.

In one of the cases (*Case* 21) the splitting or fraying of the aponeurosis above alluded to, as a consequence of over-tension, resulted in a slight protrusion at the side of the puncture in the outer pillar. This, however, became smaller under the pressure of a truss with a broad pad. Its occurrence has led me to modify the degree of tightening in the twist of the upper wire loop at the time of the operation.

The necessity of separating the fascia of the scrotum to such an extent as to allow of perfect invagination without dragging up the skin between the pillars of the superficial ring, has been before alluded to. A neglect of this precaution may cause imperfection in the adhesion and closure of this part of the canal. In some cases, the toughness of the fascia, or the presence of cicatrices from former operations, may necessitate a more careful and extensive employment of the knife than is usually required.

It is well to bear in mind, that as little rough handling and disturbance of the tissues as may be, is desirable in the manipulation necessary for placing the sutures.

In one or two cases, where thread was used for the ligature, the facility with which it is drawn too tight led to the production of small sloughs in the enclosed structures. These cases appeared afterwards to present a greater degree of bulging at the groin opposite the internal ring, in consequence of the diminished bulk and strength of the cicatrization. Latterly, I have been careful to twist the wire ligature to such a degree of tightness only as was necessary to hold firmly together the enclosed parts.

The bulging above alluded to occasionally remains after the operation, for the same reason that it exists in many cases of predisposition to rupture where no rupture has actually taken place — viz., from a deficiency in the development of the lower part of the internal oblique.

In some, I have succeeded in correcting it by placing the outer wire very high, so as to cover the deep ring, and cause adhesion between it and the superjacent aponeurosis. In the cases in which invagination of the fascia was done, the hard cord of adhesion resulting from the union of its raw surfaces can usually be felt plainly in front of the cord, and blocking up the superficial ring long after the operation. In a few instances it has been observed, after some time, to shift slowly downwards under the weight of the testis and scrotum, which had been previously tucked up by it. This occurrence appeared to indicate a less perfect closure of the superficial ring than might be desirable, and a tendency in the tissues to slide over each other. It is not to be taken, however, as an indication of the return of the rupture, since its connexion with the cicatrices in the inguinal canal causes it, as it descends, to draw the walls of the passage closer to each other, as shown by its effect upon the upper cicatrix. By the patient, or an inexperienced observer, it is apt to be mistaken for a return of the rupture. (See *Case* 43, *Appendix*.)

The condition of the patient's viscera and general health previous to the operation must be carefully examined and considered.

It is of essential importance that his powers of reparation should be vigorous and well sustained. Upon this depends much of the permanent nature and resisting power of the newly formed adhesions. The urine should be examined for albumen and phosphates, and the action of the liver upon the fæces carefully noted. An examination by the stethoscope of the heart and lungs should also be instituted.

With a view of sustaining the vigour of reparative action, I have recommended a generous diet as soon as the patient feels inclined to take it after the operation, care being taken previously to have the bowels fairly opened. If there be more suppuration than usual, bark, the mineral acids, and chloric ether may be advantageously prescribed; and stout, wine, and brandy given. The testis should be carefully supported, and, if inflamed or painful, evaporating lotions, or a leech or two, may be applied. I have, by attention to these matters, avoided, in all the cases except two or three, any disagreeable complications in the testis, cord, or scrotum during the after treatment.

In stating the reasons for the change from the use of threads to that of wire, it was mentioned that the proportion of adhesive and consolidating action is not at all measured by the degree of suppuration induced. Still less are the permanent adhesions dependent upon the violence of the local action following the operation. In the earlier cases, I was inclined to hasten matters by a tight compression and an early removal of the sutures. Although succeeding in many cases in obtaining an earlier escape from the recumbent position, one or two failures from this cause have induced me to leave the sutures "in situ" for a much longer period; which the non-irritating properties of the wire latterly employed enable me to do without inconvenience.

An attentive observation of the daily progress and results of the cases so treated has led me to believe that a slow continued action, carefully regulated by the occasional pressure of a bandage upon the wire loop, gives rise to a more solid and permanent change in the canal and sac without the occurrence of the more disagreeable symptoms, which interfere with the favourable progress of the adhesive action. The effect of this pressure upon the testis should be carefully noted. A slight enlargement is a favourable sign, as indicating a sufficiently close influence upon the cord in the hernial canal. So slightly is the testis affected in most cases, that a gonorrhœa, which happened in two cases to have been contracted just before the patient's admission, did not give rise to any great increase of irritation in the gland, and urethral injections were continued during the whole course of the treatment. (*Cases* 12 and 38.)

In three cases, the testicle was permanently affected after

the operation by a degree of atrophy. In these it was evidently originated by sloughing, from endemic causes, or from peculiarities in the apparatus employed, which the result of these cases led me to discontinue. (*Cases* 17, 25, and 52.)

The activity of the developing and reproductive powers during early life gives a much greater chance of success in the cure of rupture when the patient is young. In a child, a closure of the hernial canal, even if so slight as to serve only to keep out the bowel for some months, gives a stimulus to the development of the parts which will lead to a complete and permanent cure. In congenital cases, the operation by pins may be performed at a very tender age without any fear of bad results. I am in the habit of operating as soon as the point of the little finger can be passed into the hernial canal. The use of a small pair of pins will then close up the canal, so as to keep out the bowel effectually, the closure being maintained by the subsequent use of a proper truss. In the smaller kinds of rupture in young boys, the same operation may be done without any symptoms whatever resulting from the operation. In some cases under my care, the tongue has not even been whitened in the whole progress of the treatment, and no pain has been experienced except during the one or two dressings which were necessary for cleanliness.

In these cases, also, a careful examination of the abdomen is a necessary preliminary. Any tubercular disease, especially of the mesenteric form, strongly contra-indicates interference.

In healthy children, on the other hand, a great deal of operative manipulation will be borne by the hernial offset of the peritoneum with perfect impunity, and much can be done with the testicle and cord which cannot, without more serious symptoms, be practised when the generative organs are fully developed and active.

The radical cure of rupture is, therefore, an operation

CAUSES OF FAILURE AND DANGER. 159

especially suited for children and young lads, in whom the probabilities of a cure by the pressure of a truss may be slight, while the great restorative and plastic power of the tissues, especially about the procreative organs, seconds vigorously the efforts of the surgeon.

For ruptures occurring in adult life, the operation by wire is more suited. Except under specially favourable conditions of health, or peculiar exigency of circumstances, I do not recommend the operation at a late period in life. A proper truss may then effect all the relief that may be required. Where trusses fail altogether, however, an operation may be safely done, and so far successfully as to render a truss afterwards available and effective for support.

XII.

SUMMARY OF THE CASES.

A FAIR objection may be entertained against any operation for the cure of rupture, if many fatal cases result after its fair employment and skilful performance, or if such serious symptoms commonly arise as call for active and debilitating measures for their suppression.

That such an objection cannot be urged against the operations practised by the author, a study of the cases appended to the volume will prove. Of the cases there given, there is but one fatal. (*Case* 20.) In this, the sad result was clearly due to a disease which, though certainly following the operation three weeks afterwards, cannot be said to be otherwise than an accidental occurrence, inasmuch as it occasionally, and at certain seasons, frequently, follows the most trivial interference with the integrity of the tissues. The uncontrollable course and fatal results, by which *pyæmia*, the disease alluded to, is usually characterised, render it one of the most portentous enemies of the surgeon in his endeavours to remedy the deformities of the human frame. Objections, therefore, founded upon such a case, would apply equally to any surgical proceeding opening the skin, even to venesection, which has frequently been followed by this disease, and so proved fatal.

Of the fatal cases I have heard of in the practice of other surgeons, one was owing to erysipelas, which was present in the wards of the hospital, and another to the violence and struggles of the patient at the time of passing the needle, rendering it probable that the bowel had been included or punctured.

The severity of the symptoms following the operation will be best judged of by an inspection of the *Appendix*.

The average duration of the cases, though hardly a fair test of an operation which requires the slow action of ulceration and adhesion, under acupressure, to produce a safe and permanent result, still presents these operations in a favourable light in this respect. It must be borne in mind, however, that in many the recumbent position was rendered advisable, to avoid testing too much, at an early period, the newly-formed adhesions. In children, this can only be effected by keeping them safely in bed. With but few exceptions, the general symptoms were so slight as to call for little attention.

Of the cases in which the duration of treatment was over thirty-one days, one was kept in bed two months from burrowing of matter and profuse suppuration; one from sloughing of the hernial sac; another from sluggish action in the parts retarding consolidation. Two children were kept in the hospital by small additional operations, rendered necessary by the enormous size of the ruptures, and the large abdominal openings. The symptoms in these latter cases were throughout so slight that the tongue was scarcely whitened, the appetite being almost wholly unimpaired.

Some of the cases were kept in bed until the proper trusses could be supplied; or for reasons totally unconnected with the severity of the operation.

The average duration of the cases in bed has been about twenty-one days. Of the 60 cases given in the *Appendix* (one case being operated on on both sides), 21 were operated on by thread and compress, 27 by various modifications of the wire operation, 10 by the use of a pair of pins, and 2 by a combination of pins and wire. The number of *failures* in the 60 cases operated on is 11. *Five* of these followed the operation by ligature-thread and compress (one being a female). (*Cases* 2, 6, 14, 15, and 18.) One of these failures is accounted for by the indiscretion of the patient in returning to very heavy work, as a sailor, three months after the operation, without truss. (*Case* 18.) Another also was owing to the neglect of the patient to continue his truss when doing very laborious harvest work, such as forking hay, &c., a few months after the operation. (*Case* 6.) Under such trying circumstances as these, the adhesions resisted a fresh protrusion for a twelvemonth after the truss had been discontinued. Even at the present time (three years after) it is smaller than when operated on. Another was owing to the age of the patient at the time of the operation, and the laborious nature of his subsequent occupation, as a seaman in a collier ship. (*Case* 2.) In this case there was also some sloughing at the time of the operation. *Case* 14 is accounted for by considerable alteration having resulted in the anatomy of the parts from a previous severe operation, rendering a repetition of the operation by threads and compress also unsuccessful. In the female (*Case* 15), I found it difficult to obtain such invagination of the fascia as to enable me to reach the higher part of the inguinal canal, and to plant the sutures fairly.

Four of the 26 cases operated on by wire failed to effect a cure. In *Case* 23, this was clearly owing to the nature of the man's occupation—viz., pushing heavy trucks along with the belly, at the same time wearing no truss. In this case, the tenacity of the adhesions was demonstrated by the rupture not reappearing until four months after the truss was discarded, under the idea that the rupture was permanently cured, after wearing it four months only. In *Case* 40, the rupture was immensely large, and passed directly out of the abdominal cavity by a very large opening. The protrusion which happened after the operation, occurred after a hard day's work as a Covent-garden porter, and was only a tithe of its original size. The opening into the abdomen is at the present time much diminished, and barely admitted the point of the finger when I last saw the case. I have no doubt that a repetition of the operation, if the man will submit to it, will achieve a complete success. It will be observed that in this case, although large and direct, I performed the simpler of the operations by wire, with a view of testing its capabilities. The result, indeed, can scarcely be called a failure; although the success was not perfect, the man's condition was very much improved. *Case* 29 was a rupture attended with a very loose condition of the hernial walls, and a conformation of the groins and genitals which I have before described as partaking of a fœtal character. Such cases may be considered as difficult to cure, on account of the general laxity of the tissues, and the yielding nature of the new adhesions. In this case, I had some doubt as to having secured properly the conjoined tendon. In *Case* 36, there was unfortunately some sloughing of the tissues in the track of the ligature, which had the effect of weakening the parts after recovery. The age of the patient (fifty-seven) militated against a satisfactory result; and he wore no truss after the operation, until the hernia reappeared.

In *two* only of the *ten* cases in which the rectangular pins were used, has a failure been the result. The second case operated on (*Case* 31) was one of double congenital rupture, on both of which the operation was performed. The operation first performed was completely successful. After the other side was done, the patient, from an attack of teething, diarrhœa, and cough, became much emaciated and reduced; a tumid belly and hardened mesenteric glands revealing the existence of tubercular disease. According to the mother's account, the rupture on this side returned as a small tumour several times after a smart attack of coughing. In many examinations without truss, and under coughing and crying, I was myself not able to detect any protrusion. When last seen, however, the mother said that some months ago the tumour had reappeared to some extent several times, (after an attack of bronchitis and diarrhœa, which produced much emaciation,) but had not lately been observed. In another case (*Case* 33), the subsequent occurrence of a varicocele and an obscure swelling, to some extent rendered the case a doubtful one. When last seen, the signs of a return of the rupture were more decided.

The two cases in which pins were first applied, and afterwards a wire suture, were cases so exceptional in their enormous size, the thinness of the sac coverings, and the deficiency of the abdominal walls, that I had but small hope of a complete success, and operated at the mother's earnest request, in order, if possible, to reduce the rupture in size, so as to render a truss available. Though time is not yet, perhaps, sufficiently elapsed to enable me to pronounce with certainty upon the ultimate result under such formidable circumstances; yet in one case the result is at present so satisfactory as to render a complete cure all but certain; and in the other, a satisfactory termination is very probable.

It is important to mention, that in none of the cases operated on unsuccessfully, as regards a complete cure, has the rupture returned to the full size which they had attained before the operation. In none has the rupture been rendered irreducible, or become strangulated, and all can wear a truss with more comfort than before the operation. The facts just mentioned remove entirely one of the most important of the objections sometimes made to an operation for the radical cure of rupture.

In *six* more of the cases the result may be considered at

the present time as either doubtful, or the success only partial. In three of these, however, the ruptures were so enormous that the operation was only attempted on the understanding and with the view of rendering the use of a truss available.

One of the cases operated on by ligature-threads and compress resulted in a varicose condition of the spermatic veins, and an effusion into the fundus of the sac, which rendered the condition of the patient, when last seen, doubtful as to the ultimate result. (*Case* 5.) Having, however, met with an exactly similar state in another case which subsequently resulted in a satisfactory cure, I have hesitated in recording it as a confirmed result, either successful or unsuccessful.

In another case, after an operation by wire of an experimentary character (*Case* 22), a small protrusion afterwards took place, apparently through one of the punctures in the aponeurosis. This, however, was perfectly commanded by the pressure of a truss, which was not the case with the original rupture; and it will probably be completely closed up in time, by carefully persevering in the proper truss-pressure.

One patient (*Case* 14) was twice operated on by the ligature-thread and compress unsuccessfully, for reasons before given. Two other patients were also twice operated on for the same rupture. One was done first with the thread and compress, and afterwards with wire in the ordinary manner (*Case* 12.) The return of the rupture after the first operation was due entirely to the patient's recklessness in lifting a heavy weight three weeks after the operation, without the support of a truss. The second operation was done more than a year ago, and has been perfectly successful up to the present time. The other case (*Case* 37) was one of direct hernia, scrotal, and with a large internal opening, in a muscular man. Both operations were by wire—the first, the ordinary operation; and the last, the operation illustrated in Figure 36. After the first operation the rupture returned on making efforts to evacuate the bowels without truss. After the second operation, the case remained in a satisfactory condition up to the period of his discharge from confinement.

Two of the cases have been followed by some degree of atrophy of the testis from causes removed by a modification in the operation, and one from sloughing in the scrotum, originating from impure air. In two cases, a coexistent varicocele was completely cured by the operation, as well as the rupture. One of these varicoceles was enormously large. In two others, some atrophy of the testicle which previously existed, seemed to improve, and the gland to increase in size, after the operation.

Of the 42 cases put down as successful, 10 have not been heard of since they left the hospital cured. The rest have either been seen (and many exhibited in public from time to time), or have been heard of by the author. All the patients were earnestly enjoined, and promised to communicate directly with the author in case of any change in the condition of the parts operated on. Some of the cases present a slight degree of bulging opposite the internal ring, which results from the tenuity of the abdominal wall at that part. These might possibly have been improved by a higher position given to the outer point of suture. Such a weakness of the groin is, however, commonly met with in persons not actually affected by rupture, as we have before seen.

Since the sixty cases given, three or four more have been operated on, but so recently that no deductions can safely be drawn from them, though they have been fairly tested by coughing and walking without truss, and are in a very satisfactory condition.

Eight of the cases operated on by thread and compress were

SUMMARY OF THE CASES. 167

treated entirely without truss throughout, as an experiment to test the reality of the cure, and the proportionate share to be fairly attributed to the support given by artificial means. Of the six cases shown at the Medico-Chirurgical Society in the month of February, 1860, one has since had a decided return of the malady. (*Case* 6, before alluded to.) This and three others of these cases were among those treated entirely *without truss*. The patient first operated on, one year and eleven months before the above date, has been doing very heavy lifting work without truss up to a recent period, nearly five years after the operation.

The total result, then, of the cases given in the *Appendix* yields, without reckoning the doubtful and imperfect cases, and with a fair allowance for future casualties and imperfect records, the encouraging proportion of 65 to 70 *per cent. of successful cases*. And this, it must be borne in mind, is drawn from cases that have been taken entirely *without selection*, and as presented, good and bad, direct and oblique, many of them of a very aggravated kind, some of enormous size, and treated by operative measures some of which were more or less tentative.

XIII.

THE USE OF TRUSSES IN INGUINAL HERNIA.

It has been a matter of common observation, both to myself and other surgeons, that by the continued wearing of the trusses in ordinary use, although the rupture may be entirely retained, yet the aperture becomes gradually larger, and the protrusion, when suffered to appear, is also constantly increasing in size.

This is caused in the following manner:—The truss-pad is usually made more or less convex or conical on the surface applied to the skin, in order that, by biting into the soft parts, and making for itself a cup or depression, it may hold its place firmly over the rupture with less retaining apparatus, and give less trouble in fitting and retaining. The effect is, that the soft parts are pressed like a wedge into the already dilated inguinal openings, opening and stretching the intercolumnar fasciae and bands, and operating upon them from without exactly in the same manner as the bowel itself does from within. The chief force of the side spring is exerted in the axis of the dilated canal and hernial openings, and its upward and downward movement upon the haunch during exercise acting upon the conical pad as the centre of motion, works it continually deeper into the superficial ring and hernial canal, and gradually dilates them by a sort of invagination.

It has been seen, in treating of the pathology of inguinal hernia, that the borders of the superficial ring are everted and protruded by the rupture so as to be divergent from each other, at the same time that their fibres are opened out and thinned by distension, and eventually, in large herniæ, are lost upon the coverings of the sac. The intercolumnar fibres around, and their connecting fascia, are thus stretched out and weakened, permitting a greater separation of the parallel fibres of the aponeurosis, giving a more open texture to this structure, and diminishing its resisting powers.

If pressure by a rounded convex surface be made upon the centre of this thinned and bulging wall, its concavity is simply inverted, and the dilatation upon the sides of opening continued to the same extent as before. In many of the trusses in common use, the lower extremity of the pad is made the most convex and projecting part, and is the part most prominently forced into the external ring by the action of the side spring. Such truss-pads, especially if small, are very objectionable, a constant state of invagination of the soft parts being maintained.

After a rupture is returned by the taxis, the fingers of the operator retain it in the abdomen, not so much by pressure in the centre of the protrusion, as by grasping the sides and compressing them together by pressure upon the pillars of the superficial ring. Such a compression of the sides of the hernial canal and openings, if continued long enough, would promote the contraction of the calibre of the canal, and give a much greater chance of an ultimate closure. In small and incipient cases, such an application of the pressure may be fairly expected to produce a cure without further interference. Undoubted cases are on record in which the pressure of truss brought to bear, perhaps accidentally, upon the proper place, has been followed by this satisfactory termination. The chief reason why such good results do not more commonly ensue, lies in the faulty construction of the truss-pads ordinarily worn.

Much more objectionable are these trusses after the operation for a radical cure has been performed. The pressure of the highly convex surface falls exactly upon the adhesions which have been obtained in the ring and canal, binding together their sides. When the resistance of the pillars of the superficial ring is thus weakened and overcome, the conical pressure falls upon the conjoined tendon and posterior wall, forcing them backwards in the direction in which the contractions of the rectus muscle tend to draw them, and producing by continued pressure absorption of the adhesions within the canal. Continued observation of the effect of such trusses upon the cases in which the rupture has returned after operation, has convinced the author that, in many of them, it has been the main cause of the reappearance of the hernia. These effects are much accelerated by the movements of the side spring working over the hip during exercise, and boring the pad into the rings, especially if the spring be strong enough to compress instead of merely supporting. The adhesions are thereby gradually loosened, and the sliding of the transplanted structures over each other promoted.

The above considerations have led me to the construction of truss-pads for the different forms of rupture, both before and after operation, more in accordance with the necessities of the hernial condition and the mechanical relations of the weakened structures. These will, I think, be found to obviate the faults just pointed out.

Truss for Oblique Inguinal Hernia.—The essential peculiarities of this truss-pad are, that it is an elongated ovoid with a flat surface, wider above than below, and with a chink cut clean out at the lower half of its long axis, making it resemble altogether a horse-shoe made oblique, with the extremities well rounded off, and one a little shorter than the other. (See Fig. 48.)

For ordinary cases and hospital purposes generally, no better material can be got than boxwood of which to construct the pad. It should be about five-eighths of an inch thick, and have the ends and corners well rounded off, and may be left quite uncovered. It is attached by a screw to the end of an ordinary side-spring at about its geometric centre, with a couple of studs at the ends to fix an under strap to. The top of the screw being provided with a brass stud grooved across, serves both to fix the body strap and to adjust the angle of connexion with the spring. The size must be regulated by that of the rupture; the larger it is the better, consistently with comfort to the wearer. It should cover all the weak parts in the

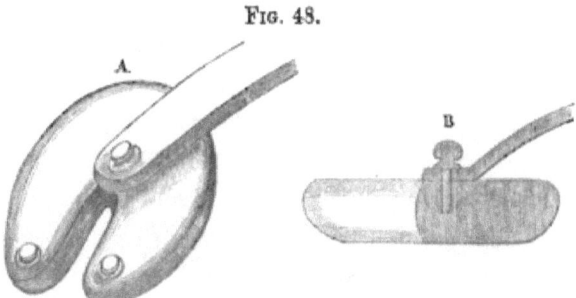

Fig. 48.

The Horse-shoe Pad.—A. Superior view. B. Section.

groin, and fit easily into the curve of the groove between the thigh and abdomen. The size most generally useful is from 2 to 2½ inches long by 1¼ to 1¾ inch broad. The upper end of the oval is broader than the lower, so as to cover and support the deep ring. The fissure is to be placed over the opening of the superficial ring, or over the adhesions which block it up after operation. At the lower part of the chink the spermatic cord emerges, and thus escapes the hard pressure of the pad. The spine of the pubis should also be felt between the ends of the horse-shoe. The inner horn of the horse-shoe is larger and longer, and is intended to be placed upon and compress the inner pillar of the superficial ring. The outer end, about half an inch shorter, tapers gradually to the rounded point, corresponding in shape to the triangular outer pillar of the ring upon which it fits, its lower border lying upon, and shaped to, the groove of the groin. The proportionate size of the horns and intervening slit may be increased or diminished according to the width of the pillars of the ring in each individual case. If an extraordinary projection of the pubic crest and spine require it, the lower ends may be well sloped off in the downward and inward direction, care being taken to preserve the general flatness of the surfaces when taken laterally or transversely. The screw which fixes the side spring should be placed close to the upper end of the chink, as near as possible to the geometric centre of the pad, so as to bring its pressure to bear fairly at the middle of the inguinal passage.

The double bearing of the horse-shoe ends of the pad, and its accurate adaptation to the fold in the groin, give a great steadiness to its hold upon the surface. A slight projection of a fold of skin into the fissure tends still further to secure it against shifting. The limited diameter of the chink, and the tension of the skin upon which it rests, prevent any protrusion of the hernia, or any strain upon the adhesions at the superficial ring; while the spermatic cord and pubic spine escape pressure between the rounded extremities. The studs placed at the ends of the horns enable increased pressure to be made upon them, when desirable, by the tightening of the under strap.

For cases more difficult to control, either from the size of the superficial or deep hernial openings, or from the protuberant abdomen and deep groin resulting from a great obliquity of the pelvis, and an increase of the lumbar curve, described in the section upon the pathology of rupture, I have contrived a simple combination of *pad springs* and *lever*, which will transmit the pressure of the hip spring, or body strap, to any part of the compressing surface where the rupture may have a tendency especially to escape. In some cases, this occurs towards the internal pillar; in others, behind the external pillar.

As the basis of this form of truss pad is employed a thin plate of steel or German silver, curved up round the edges into a sort of dish-shape, with a border half an inch deep.

Placed in the hollow of the plate, and working in each of the horns of the pad, is a *double C-spring*, the limbs of which are connected at the upper end by a broader part, so as somewhat to resemble in shape the letter Y. (See Fig. 49, A and B, *a*.)

FIG. 49.

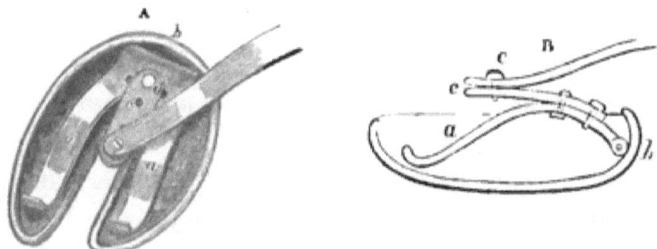

The Horse-shoe Lever Spring Pad.—A. Superior view. B. Section.

The upper end of the united springs is held to the border of the plate by a common hinge (*b*), upon which the spring can be turned back to adjust the attachment of the side spring, or strap. The curve and strength of the springs may be increased or diminished according to the degree of retaining force required. It is necessary, however, to proportion them carefully to that of the hip-spring, when this is used. The elastic reaction of the pad springs, taken together, should exactly equal that of the body spring or belt, when either of these is employed.

The shank-piece connecting the springs is prolonged between them so as to overlap the centre of the pad, forming a *lever* upon which the side spring or strap may be fixed by a simple screw and nut, which should be screwed tightly down. To give a power of adjusting the point of pressure of the side spring, six holes, made in the lever-plate in the shape of a square, for the reception of the screw and nut, will enable the maker to shift the pressure higher, or more laterally, as the case may require. This method of fixing is seen at *c*, in Fig. 51. A more nice and ready adjustment may, however, be obtained, (as given in Fig. 49,) by having an additional plate carrying the side spring, moveable upon the lower end of the lever, round a pin near the centre (*c*), and fixed in the required position by a hand-screw above (*d*), fastening into holes in the spring plate. By this means the pressure can be increased or diminished upon one limb or other of the horse-shoe pad. When fitted, the end of the body strap buttons upon the screw-head of the side spring. If an under strap is necessary, it may also be fastened upon the same point, or studs fixed at the ends of the horse-shoe may be used for that purpose, if the case seems to require it. One or other of these, or both, may be fastened to the under strap. The compressing surface of the metal pad may be covered by a thin layer of vulcanized india-rubber, which passes across the fissure between the ends, so as to restrain by its elasticity the projection of the skin between them. The whole is covered by wash-leather, or other soft substance, made into a sort of case or bag.

In a few cases I have employed a double spring of the shape of the letter H, working under braces at the upper and lower ends of the pad plate. The attachment of the hip spring is effected by a screw sliding along a slit in the central transverse bar. This arrangement acts very well, but does not present the regulating capabilities of that last described, which possesses also the additional advantage of a lever pressure upon the lower part of the pad.

The pad may be fitted to a spring over the hip on the same side as the rupture, as in the ordinary manner; or, what in some cases is better, to a spring over the opposite hip, as in Salmon and Ody's plan;—in this way no under strap is usually required.

The spring pad just described will in many cases be equally effective under a simple circular body belt without side spring, combined with an under strap, in the manner of the *mocmain* truss. The hollow curve of the groin may be filled up by an increased central elevation given to the springs, so as to render effective the circular pressure of the body strap, at the same time accommodating the pressure to the movements of the abdomen, and bringing an equable and steady bearing, which will not yield readily to a tilting lift of the hernial protrusion, upon the circumference and extremities of the compressing surface of the pad.

When applied, the finger should be placed upon the projecting spine of the pubis, the cleft between the forked ends of the pad should be then placed directly above the finger, and placed nearly parallel to the groove of the groin. The adjusting screw should then be fixed immovably tight, giving the pad the proper degree of obliquity. In corpulent persons, a considerable inward twist should be given to the hip spring near its extremity, so as to permit the pad to lie flat upon the surface. An under strap, properly fixed, will then keep up the proper angle, and so produce a flat pressure upon the hernial surface.

By the use of this pad we have the boundaries of the hernial canal and openings supported and compressed by a flat surface, tending to resist outward and lateral dilatation in exactly the same manner as the fingers of the surgeon. There is no dilating pressure in the axis of the canal whatever, and no invagination of the soft parts, as in the use of the convex truss-pads. The whole of the canal is equably compressed, and the deep ring supported by the upper part of the truss. The difficulties of maintaining the position upon the proper site which are usually experienced in the use of a perfectly flat pad, are met by the double bearing of the horse-shoe forks, and the anatomical outline given to the borders. An extended experience acquired with this pad since its adoption has convinced me that it retains its position as well as any pad before invented, and has not, as nearly all have, the injurious property of increasing the size of the hernial openings by invaginating the hernial coverings into the rings by convex pressure. In small and incipient cases, the patient may expect from its use the utmost benefit that pressure can effect towards producing a radical cure without operation. Adapted to the anatomical peculiarities of the weakened part, its mechanical arrangement tends, under all positions of the body, to exclude the bowel from the sac, and to close up the sides of the hernial canal. In two cases of oblique hernia associated with a thickened and varicose condition of the spermatic cord, now under my care, the horse-shoe pad has answered admirably in easing off the pressure from the cord, by allowing it to lie opposite to the chink (made wider than usual for that purpose). At the same time the rupture is perfectly retained. In one case, the inflamed and tender condition of the cord had been entirely produced by wearing an ill-fitting convex truss.

Truss for Direct Inguinal Hernia.—In this form of rupture, the neck of the sac and sides of the opening, being short and limited, a somewhat different shape and adjustment of the pad is required.

The shape of the pad best adapted here for making a closing pressure upon the hernial ring is an ovate, with flattened surfaces; with a somewhat larger curve at the lower border, and a more contracted one at the upper. (Fig. 50.) The size most generally required will be about 2 inches to $2\frac{1}{2}$ inches in its widest diameter. It is perforated in the middle by an opening which occupies the middle third of its diameter.

For ordinary and hospital uses it may be made of boxwood, like the last, and secured to a side-spring by a screw fastening on to the centre of a transverse piece of metal bridging over the central opening, as seen in the figure. To a couple of studs placed near the middle of the lower part of the ring, the under strap is buttoned in the ordinary way.

For particular and difficult cases it may be desirable to have the power of adjusting the pressure over the borders, so as to increase it at any part where the rupture shows a tendency to emerge.

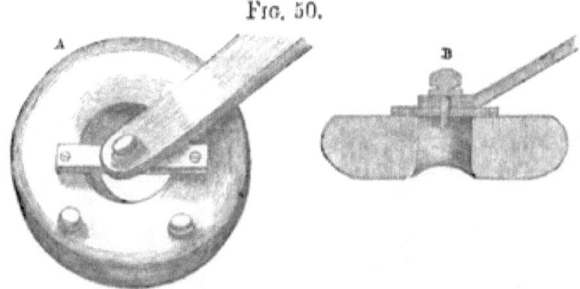

Fig. 50.

The Ovoid Ring Pad.—A. Superior view. B. Section.

For this purpose the basis of the pad should be, as before described, a thin plate of steel, or German silver, beat up at the edges into a dish-shape, and covered on the bearing surface with a layer of vulcanized india-rubber stretching over the central opening. The pad-springs for distributing the pressure in this case are arranged in the same manner as those described with the horse-shoe pad. The springs have, however, a vertical instead of an oblique direction, their extremities bearing upon the lower part of the ring. In Fig. 51, the adjustment of the attachment of the side-spring, or body-belt, to the lever-plate, is drawn after a model of the more simple arrangement, before described. This method of fixing is equally applicable to all the forms of spring pads herein described. The elevation of the springs at the highest point of the curve will be rather above the level of the curved sides of the disc. (Fig. 51, B.) By the adjustment of the screw (*e*) in the openings of the lever-plate (*c*), the force of the side-spring may be increased upon either side of the lower circumference of the ring. The strength of the springs must be regulated by that of the side-spring, which may be worn over either hip. The whole is covered and secured from the effects of damp and dirt by an investment of wash-leather.

The circular flat bearing of this pad renders it less liable to

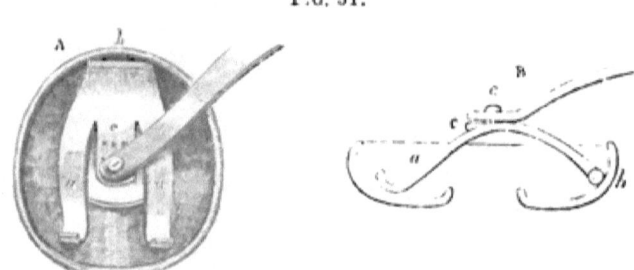

Fig. 51.

The Ovoid Lever-spring Ring Pad.—A. Superior view. B. Section.

be displaced than the convex or conical pad. The conformation of its margin to the shape of the groove in the groin contributes also to sustain it in the proper position; while the spermatic cord escapes its pressure at the outer side of its bearing on the upper border of the pubis. Firm pressure is thus made upon the margins of the hernial opening; no invagination of the integuments is possible, and a closing instead of a dilating action upon the hernial passage is effected. After the operation for a radical cure the adhesions lying towards the centre of the region thus escape that injurious pressure which tends to stretch them, and induce their absorption.

The Double Pad and Spring Truss.—In cases of inguinal

rupture of either kind where, from the laxity of the pillars of the superficial ring and of the tendinous structures in the groin generally, it is difficult to keep the sides of the canal in close apposition, success may be better attained by the use of the double spring and pad truss, of which an engraving is here subjoined. (Fig. 52.) This, as well as the other kinds of truss, is made by Mr. Matthews, of Portugal-street, from the author's designs. In this truss the pad is made in two separate portions, which, when applied to each other, make a compound pad with an oval outline. The inner portion (*a*) is rather

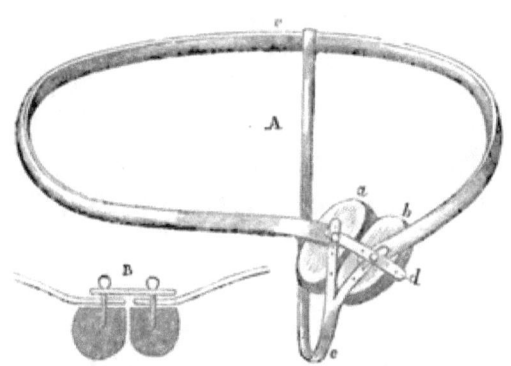

Fig. 52.

A. Anterior view of Truss.   B. Section of Pad.

longer, and fits upon the internal pillar of the ring: the outer one (*b*) rests in the groove of the groin, exerting its pressure upon Poupart's ligament and the outer pillar of the ring. A single spring (*c*) passes round the hips and back. Its longer end is fixed by a revolving joint upon the inner segment of the pad, and the opposite or shorter end upon the outer. This spring may be made to adjust at the central or back part, like that of a truss for double rupture. A strong elastic band (*d*) fixes the two segments of the pad firmly together, when applied, by buttoning upon the two studs which fix the ends of the body-spring. An under strap (*e*), fastened by two divisions upon the pair of studs before mentioned, or, if the case requires more pressure at the lower part, upon two other studs fixed into the lower ends of the twin half-pads. The half-pads may be made of box-wood, or of some softer material, such as vulcanized india-rubber, fixed upon a basis of metal. The surfaces applied to the skin are convex in transverse section (as seen in Fig. B), but nearly straight in the longitudinal direction.

In this truss we have a separate pressure applied upon each of the moveable halves of the pad, and directed upon each of the pillars of the ring independently, in such a manner as to press them towards each other, and to close up the hernial canal towards its axial line. No invaginating effect can be produced, the double bearing of the halves leaving a chink in the middle line. The pressure can be increased upon either of the pillars independently, by increasing the power of the corresponding extremity of the body-spring.

PART II.

FEMORAL OR CRURAL HERNIA.

The much less frequency of the occurrence of this form of rupture, and the habits of the persons usually affected by it (females), render operative interference for its radical cure less frequently demanded than in the case of the inguinal variety.

It may also be said that the proximity of the great vessels of the leg and their accompanying lymphatics to the site of the rupture render necessary a great amount of anatomical skill and manipulative steadiness in the operator, and expose the patient to a somewhat greater amount of risk from the effects of ulceration or purulent absorption in the after-progress of the case. At the same time, it must be admitted that it is more difficult to obtain for this kind of rupture an efficient and properly fitting truss, that patients in consequence often suffer much for want of support, are more subjected to the dangers of strangulation, while a cure by pressure is rarely obtained. The light nature of the occupations of females generally renders the pressure of a properly fitting truss sufficiently effective to retain the rupture, but cases are not uncommon in which a laborious occupation under these circumstances hourly endangers the life of the individual, as in basket-women, milk-women, and others. These employments involve a great amount of strain upon the parts, often rendering trusses useless in keeping up the rupture, and supplying our hospitals with the chief part of the cases of strangulated hernia.

Under these circumstances so dangerously affecting persons of otherwise robust health, I conceive it to be entirely justifiable to attempt to cure or diminish the size of a large femoral rupture by operation. Although I have not hitherto had occasion to perform such an operation on the living subject, I have been in the habit for some years of exhibiting to the students at King's College the way in which it may be best effected, and with the greatest amount of safety to the patient.

To understand properly the principles upon which the application of operative methods for the radical cure, or of truss-pressure as a palliative or radical remedy, should be based; as well as those of a proper application of the taxis or operative interference under conditions of strangulation;—it is of the greatest importance to obtain an accurate knowledge of the anatomy and pathology, and of the means of recognising this dangerous form of rupture.

A brief description of the parts, succeeded by a review of their modifications under hernial conditions, and from other pathological changes which simulate rupture, will facilitate a clear conception of the steps of the operation above alluded to; and also of the action of the truss upon the hernial openings, of the probabilities of a radical cure by either of these methods, and of the proper direction to be given to efforts towards reduction by the hand.

XIV.

ANATOMY OF THE PARTS CONCERNED IN CRURAL HERNIA.

BELOW the inverted curve of Poupart's ligament, the front surface of the thigh presents a slight hollow or groove towards the inner side, having an obscure triangular form with the base upwards. The outer side is bounded by the mass of extensor muscles, and more particularly by the sartorius; the inner, by the sloping surface and raised border of the long adductor and edge of the gracilis; the apex below is produced by the approximation of these muscles to each other at the junction of the middle and upper thirds of the thigh; and the base above, by the structures of the abdominal wall. The floor of the space slopes from each side towards the centre, which is placed just below, and in front of the hip joint, and it lodges, at its deepest part, the femoral vessels and their branches. The outer part of the floor is formed by the insertion of the psoas and iliacus muscles, between which is placed each of the branches of the anterior crural nerve. The inner part is formed above by the pectineus muscle, with a small portion of the adductor brevis; and is continued downwards and inwards by the sloping surface of the adductor longus to the raised inner border. The space is called, after the great Italian surgeon and anatomist, *Scarpa's triangle*. To the inner side of this triangular space, near the base or upper part, lies the region through which a femoral rupture makes its way to the surface.

The *superficial fascia* in the crural portion of the groin, like that before described in the upper or inguinal region, is susceptible of division by the scalpel into two layers, within and between which lie the superficial vessels, nerves, and lymphatics. The superficial layer usually contains much fat;—it is continuous with that covering the abdomen, and is but loosely attached to Poupart's ligament. (See Fig. 1, *a*, page 14.) The deeper layer contains less fat, is thin, loose, and easily torn. By its deep surface it is loosely connected with the subjacent structures, except at Poupart's ligament, and at the margins of the saphenous opening in the deep aponeurotic fascia, where it is more closely adherent to the underlying parts. It invests the superficial glands, vessels, and nerves with an areolar envelope; and is perforated by many oblique openings for the passage of these structures through it. (Fig. 1, *b*.)

The superficial vessels of the part are those described in the upper inguinal region,—viz., the *superficial circumflex iliac*; the *superficial epigastric*; and the *superficial external pudic*; all branches of the common femoral artery at this place. They perforate, at different places, the deep fascia, or pass through the opening above mentioned, and diverge to their respective destinations. The *circumflex iliac* (*d*) is directed towards the anterior superior iliac spine, near which it crosses Poupart's ligament. The *superficial epigastric* (*e*) ascends slightly inwards towards the navel, crossing the same ligament about its centre a little internal to the position of the deep abdominal ring. It supplies the most of the superficial lymphatic glands. The *superficial external pudic* (*f*), passes across the inner end of the ligament, crossing the spermatic cord as it emerges from the superficial ring, to be distributed to the *mons veneris*. Besides these, there is a small branch, the *inferior external pudic*, lying under the deep fascia internally, perforating it close up to its attachment to the pubis, and crossing behind or in front of the cord to reach the pudendum. These branches are subject to many irregularities. Either of the two last may be so enlarged as to give off the artery to the dorsum of the penis. Any of them may be so small as to be scarcely observable. The veins corresponding to them all open downwards into the large trunk next described, just before it pierces the deep fascia.

The *internal great,* or *long saphena vein* (*l*), after reaching the knee, lies along the posterior border of the sartorius muscle, and in the upper part of the thigh is placed upon the deep fascia in the groove between the adductor and extensor muscles. When within one-and-a-half to two inches from Poupart's ligament, it passes through the lower part of a considerable opening in the deep fascia to join the femoral vein. Besides the small veins before mentioned, it receives a considerable tributary from the inner and back part of the thigh. In some subjects the saphena vein has a more posterior position, and its largest tributary comes from the front of the thigh. In others, these two trunks are connected just above the knee joint, and appear as a double saphena.

The lymphatics of this region are very numerous. The ducts lie chiefly along the inner side of the saphena vein, but many also on the other side (*m*). They are provided with numerous glands of an oval shape, arranged with their long diameters in the vertical direction. These are connected above with the set of glands before described, as lying upon and parallel to Poupart's ligament. The ducts of all these superficial glands pass through the same opening in the fascia lata which transmits the saphena vein.

The nerves of this region are as follow:—The *internal* and *middle cutaneous* branches of the *anterior crural;* the *ilio-inguinal; crural branch* of *genito-crural,* and *external cutaneous;* —all from the lumbar plexus. The *ilio-inguinal* (*m*) escapes through the fascia covering the superficial abdominal ring, gives off small twigs to the scrotum and mons veneris; then turning outward and downward over Poupart's ligament, it is finally distributed to the integuments about the saphenous opening. The *internal cutaneous* passes downwards and inwards across the front of the femoral vessels, under the deep fascia, giving off many twigs to the integuments and sheath of the vessels, and finally perforates the deep fascia below the middle of the thigh to supply the skin on the inner side about the knee.

The *middle and external cutaneous* nerves, with the branch of the *genito-crural*, are distributed to the integuments on the front and outer surfaces, and do not bear any noteworthy relation to the parts of crural hernia. (See Fig. 53, page 188, *a*.)

The *deep fascia* of the thigh, or *fascia lata*, is a strong and aponeurotic-looking layer of white fibrous tissue, of which the fibres are interwoven almost like a piece of cloth, the strongest passing downwards and inwards. In the groin it is arranged in two portions, named from their respective attachments. The outer or *iliac portion* is connected with the crest and superior spine of the ilium, and is closely attached to and blended with the lower part of the tendinous aponeurosis of the external oblique forming the crural arch or Poupart's ligament (*d*). With the insertion of this ligament, it is attached also to the spine and pectineal line of the pubis, upon which it is blended in a common insertion with a triangular, deflected portion of the same ligament, called *Gimbernat's* (*c*). This part of the fascia lata is placed in front of the femoral vessels, to the sheath of which it is pretty closely adherent. A falciform portion of it also crosses, immediately below Poupart's ligament, the channel by which crural rupture passes down on to the inner side of these vessels, assuming thus a very important relation to it. This part (*f*) forms, by a curved edge, the upper and outer borders of the opening which transmits the saphena vein and lymphatics to the deeper vessels.

The inner or *pubic portion* of the fascia lata (*e*) is attached to the united branches of the pubis and ischium which form part of the inferior outlet of the pelvis, where it is connected with the perineal fascia and crus penis. Above this point it is attached to the front of the symphysis and angle of the pubis, and stretching upwards, outwards, and backwards, following the border of the superior pubic ramus, it passes *behind* the femoral vessels to become attached to the pectineal line of the pubis, joining at that place with Gimbernat's ligament and the falciform process of the iliac portion, and becoming connected externally with the *fascia iliaca*, which has been described as forming the posterior part of the sheath of the femoral vessels, and with the capsule of the hip joint. In this manner the femoral vessels are placed between the two portions of fascia lata which are separated at this part by the passage of the saphena vein and lymphatic ducts.

The opening thus formed is called the *saphenous opening*. (See Fig. 1, page 14, *o*.) It is oval in shape, with its long diameter in the direction of the thigh, and the axis of its plane directed obliquely forwards, downwards, and inwards; lying over the crural canal, the femoral vein, and termination of the saphena. It is covered in by a tolerably firm network of fibrous tissue, which, when freed from the pellets of fat which are involved in it, presents the peculiar appearance of a large number of oval openings for the transmission of the numerous lymphatic ducts. Hence it is called the *cribriform fascia*. This is connected externally with the falciform margin of the iliac portion of the fascia lata, and internally with the sloping surface of the pubic portion. Below, it affords a funnel-shaped investment to the saphena vein. Superficially, it is closely connected with the deep layer of the superficial fascia investing the lymphatics; and by its deep surface with the sheath covering the femoral vessels derived from the fascia transversalis. When divested of the cribriform fascia, the *saphenous opening* is seen to present, externally and above, the crescentic margin of the iliac portion of fascia lata. This is called the *falciform process of Burns*. The upper part forms a horn-shaped process—*superior cornu*, which, from its important anterior relation to the femoral vessels and crural canal, was especially noticed by Hey, and has been called the *femoral or Hey's ligament*. This is attached internally to the spine of the pubis, and blends with Gimbernat's ligament above, and with the pubic portion

Fig. 53.

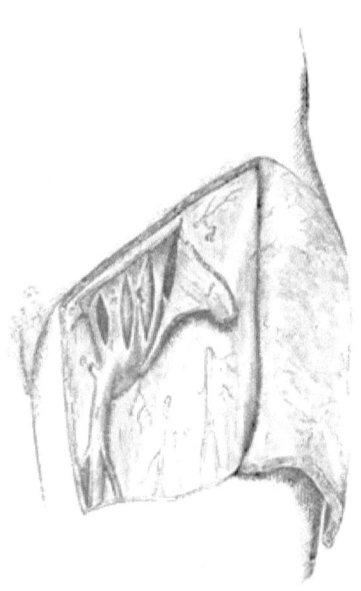

Deep dissection of the parts of crural rupture. The upper corner of the saphenous opening is separated from Gimbernat's and Poupart's ligaments and turned back, showing the three compartments of the crural sheath opened by vertical slits.

of fascia lata below its point of insertion. In Figure 53, this ligament (*f*) is seen turned back, along with the falciform margin of the saphenous opening, to show the femoral sheath placed behind it. The lower part of the falciform process—*inferior cornu*, passes under the termination of the saphena vein, and becomes lost on the surface of the pubic portion of the fascia lata at the deepest part of the adductor groove.

On removing the iliac portion of fascia lata, the sheath of the femoral vessels can be observed to emerge from behind the centre of Poupart's ligament, as a funnel-shaped investment of dense fascia. Above, it is continuous with the *fascia transversalis* anteriorly, and with the *fascia iliaca* posteriorly; the two uniting round the vessels to form a loose sheath, which encloses also, to the inner side of the vessels, the channel through which the lymphatics pass upwards into the abdomen. The sheath is intimately connected anteriorly with the deeper portion of Poupart's ligament, at which point a strong band of fibres, placed between and connecting them, forms an *arch* over the vessels and crural passage, which reaches from the psoas muscle externally, to Gimbernat's ligament and the pubic insertion of the conjoined tendon internally. This band of fibres is more marked in some subjects than in others, and is important as being usually the seat of strangulation in femoral rupture. It has been called by Cooper the *deep crural arch*. The sheath will be seen to occupy the interval between the psoas muscle externally and the edge of Gimbernat's ligament internally, reaching to within an inch of the spine of the pubis. The latter bony tubercle, usually evident enough to a careful touch in the living subject, becomes, from its position between the superficial abdominal and the crural rings, an useful guide to the distinction between the two varieties of rupture. The centre of the superficial abdominal ring is directly above the spine, and that of the crural ring from an inch to an inch and a-half external to it. The distance between the centre of the crural ring and that of each of the abdominal rings is about equal, and usually measures rather more than an inch, when in a normal condition, in the male. In the female, this distance is usually nearly half an inch greater. Externally, the femoral sheath seems to be prolonged over the psoas muscle, to form a covering for the anterior crural nerve as it emerges behind Poupart's ligament, and is finally lost in the fascia which invests the iliacus muscle to its insertion. If the latter fascia be lifted up from the surface of the muscle, it will be ascertained to be an offset from the united fasciæ iliaca and transversalis; which, after lining the posterior and anterior walls of the iliac region, form internally the sheath of the vessels, and pass down externally to become attached to the capsular ligament of the hip-joint and pubic portion of fascia lata, after uniting together at Poupart's ligament to form a "cul de sac," which closes up the fissure between that ligament and the psoas and iliacus muscles, and prevents the escape of hernia at this part. The crural branch of the genito-crural and the middle cutaneous of the anterior crural may be seen to perforate just outside the vessels.

If three parallel longitudinal cuts be now made at equal distances into the sheath of the vessels, the outer one will expose the artery; the middle one, the vein; and the inner one reveals a tubular compartment or canal, which transmits the bundle of lymphatic vessels passing into the abdomen behind Poupart's ligament (Fig. 53, *k*). A firm longitudinal septum on each side of the vein divides its compartment from those lodging the artery and the lymphatics. These septa, on further careful dissection, will be found to be derived from the proper sheath of the external iliac vessels, as they lie under the peritoneum; and they become blended with the additional funnel-shaped sheath from the fasciæ transversalis and iliaca at Poupart's ligament.

The interval between the femoral vein externally and the edge of Gimbernat's ligament internally is called the *crural ring*. In front of it is the deep crural arch, forming in appearance the deeper fibres of Poupart's ligament. Behind it is felt the pectineal line of the pubis, with the origin of the pectineus muscle, and the attachment of the pubic portion of fascia lata upon it. The fibres of *Gimbernat's ligament* may then be seen, as well as felt, to pass backwards from Poupart's ligament in an acute curve, or free border, which is prolonged outwards at its insertion into the bone so as to reach considerably along the pectineal line behind the crural ring. Here, by its union with the pubic portion of fascia lata, it sometimes forms a prominent ridge, which it has been proposed to divide by cutting upon the bone in cases of strangulated crural hernia. (See Fig. 60, *a*, page 203.)

A clear recognition of these relations of the crural ring is of great importance in the treatment of crural rupture. The ring is lined by the inner division of the funnel-shaped sheath; and through it pass the superficial inguinal and crural lymphatics. Stretching transversely across the ring, inside the sheath, is a thin layer of fascia containing fat, and investing a lymphatic gland, which transmits through its substance some of the ducts. (See Fig. 3, *k*, page 25.) This fascia was described by Cloquet under the name of the *septum crurale*. It is really derived from the subperitoneal fascia, being connected with the upper opening of the funnel-shaped sheath and with the longitudinal septum. It is rarely of sufficient thickness to form a distinct covering for crural hernia, and can have but little effect in preventing a protrusion. The *aspect* or *axis* of the crural ring in the erect posture is *downwards*, with but a slight inclination forwards and outwards. The principal direction is given by the great obliquity of the pelvis, jointly with the advanced position of Poupart's ligament in reference to the superior branch of the pubis. In the treatment of this rupture by the taxis, by operation, or by the application of trusses, it is very essential to recognise the true direction of the crural opening.

The term *crural canal* is given to the short tubular passage from the crural ring to the saphenous opening, which constitutes respectively its upper and lower terminations. It is bounded on the outer side by the femoral vein, covered by its inner longitudinal septum; on the inner, by the insertion of Hey's or the femoral ligament into the pubic portion of fascia lata, and by the inner part of the funnel-shaped sheath; in front is placed the same ligament and the front of the sheath; and behind is the superior branch of the pubis, covered by the origin of the pectineus muscle, upon which lie the pubic portion of fascia lata and the back part of the crural sheath.

The entire shape of the crural canal may be compared to an oblique slice of a cylinder (Fig 54, A) with the short side in front (c), the long side behind (d), the crural ring forming the top section (a), and the saphenous opening the more oblique lower section (b). The anterior wall (c) is represented by *Hey's* or the *femoral* ligament, which varies in width in different subjects from half to three-quarters of an inch, and constitutes with Poupart's ligament the *only* surface upon which pressure can be effectively made to keep a hernia within the crural ring. The posterior wall (d) is supported by the outer part of the pectineus muscle arising from, and resting upon, the superior branch of the pubis at the inner side of the hip-joint.

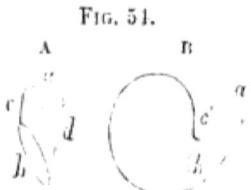

Fig. 54.

Normally, the crural canal contains only the lymphatic ducts and one or two small glands, with some fat. In crural hernia, it invests and moulds the form, and determines the direction of the neck of the sac. (Fig. B, *a* to *b*.)

The *vessels* of a size important to the operator, which are usually placed in the neighbourhood of the saphenous opening and crural canal, and comparatively superficial, are the femoral vein externally, and the saphena inferiorly. The position of

ANATOMY. 193

these can usually be recognised and avoided. Close behind the femoral vein, on its way inwards, backwards, and downwards to the inside of the hip joint, and passing close to the outer border of the pectineus muscle, is placed the *internal circumflex* branch of the *profunda*. In several instances I have found this artery giving off the *inferior external pudic* branch, which then passes across between the pectineus muscle and the fascia covering it. The precautions necessary for avoiding injury to the femoral

Fig. 55.

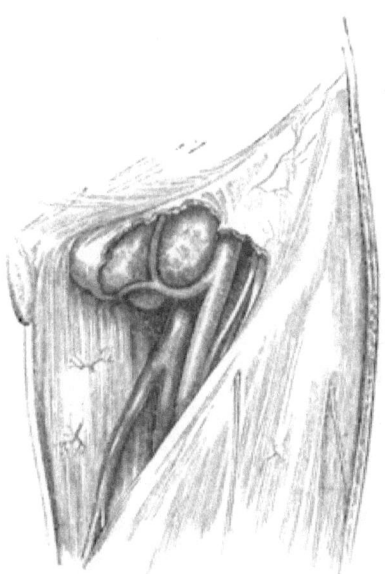

vein will usually protect this artery even in its more common irregularities.

The anterior boundary of the crural ring has, in the male, an important relation to the spermatic cord, which lies in the inguinal canal just above Poupart's ligament, crossing that structure just outside the pubic spine. In operating upon a crural rupture this must carefully be allowed for.

The irregularities of the arteries in this important region,

o

194   ON CRURAL HERNIA.

which might be a source of inconvenience in operating, are not very common. One of the most so is the origin of the *deep epigastric* artery from the *common* or *deep femoral*. A drawing of the former irregularity, taken from a dissected female subject, is copied in Fig. 55. It is the more interesting in being co-existent with a crural rupture, across the front of which the irregular artery lies. The *obturator* artery also arises from the abnormal *epigastric* after it has passed across the femoral

FIG. 56.

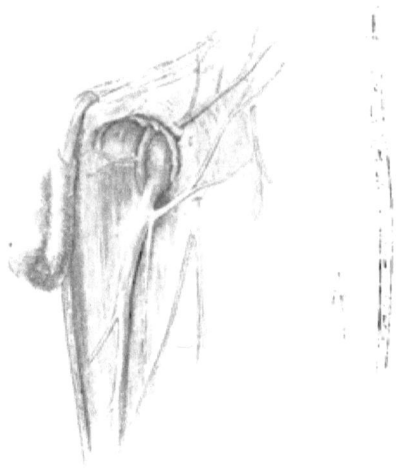

vein, and reached the middle of the hernial sac at its lower side. The sac will be seen to have pushed down before it the abnormal vessels, so as to place them in a position very embarrassing in any operation upon the rupture. The vessels would, however, be rendered tolerably apparent in such a condition by their pulsation, especially when the integuments were divided. This irregularity was present on both sides of the subject. As seen on the ruptured side in the figure, it illustrates very well the tendency of this abnormal position of the artery to resist the formation of a crural hernia, as supposed by Sir Astley Cooper. (*Op. cit.* page 253.)

In a male subject, from which Fig. 56 was taken, the *internal circumflex* branch arises high from the *common femoral*, and crosses superficial to the femoral vein, lying within the sheath, and to the outer side of and below the inner or lymphatic compartment. It gives off in its course the inferior external pudic.

Fig. 57.

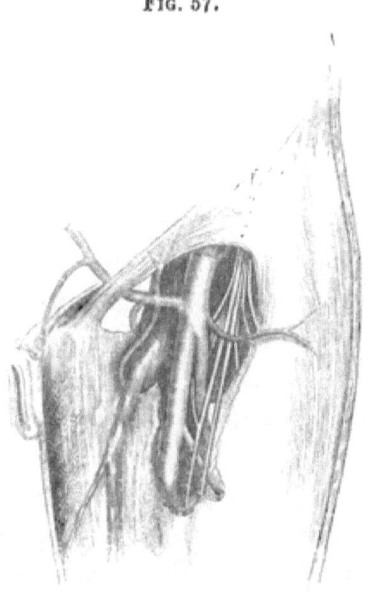

In a third subject, a female, the *internal circumflex, epigastric*, and *obturator* branches arise by a common trunk from the commencement of the *common femoral* artery. (Fig. 57.) This crosses in front of the femoral vein just below Poupart's ligament. The abnormal circumflex and obturator will be seen both to hug closely the femoral vein on its inner side, and would necessarily be placed to the outer side of a crural rupture, which might have been formed in the canal. They would thus be rendered safe by the precautions taken to avoid the vein. The same may be said of the abnormal circumflex in the preceding figure.

By examining the *crural canal* from its *upper* or *peritoneal* aspect, a relative view of the position of the inguinal and crural openings will enable us better to estimate the arrangements which predispose to one or other of these forms of hernia, the operation of the means of cure, and the deeper dangers which waylay the operator in this region.

On opening into the peritoneal cavity, and drawing the viscera to the opposite side, the serous membrane lining the inner surface of the groin will be seen to present three slight hollows or depressions, placed with reference to each other as the angles of an inverted isosceles triangle. Two of them are above and one below, a smooth oblique elevation or broad line indicating the position of Poupart's ligament. The more external one is also indicated by an indistinct cicatrix, showing the position of the internal abdominal ring. The inner one is shallower and less distinct, and occupies the interval between the epigastric and obliterated hypogastric vessels in the position of the *triangle of Hesselbach*. The one below Poupart's ligament is placed over the crural ring, at the distance of about an inch from each of the upper ones. Its presence reveals the first tendency to the protrusion of a crural rupture.

On removing the peritoneum, the two upper will be seen to be closed in by the fascia transversalis, and the lower by the septum crurale. This septum is seen to be closely connected with the sheath of the femoral vessels externally; and less intimately with the margins of Poupart's and Gimbernat's ligaments in front and internally, and with the pubic ramus behind. (See *k*, Fig. 3, page 25.)

The *epigastric* artery (*b*) on its way from the external iliac lies in the sub-serous tissue, at first behind Poupart's ligament, and placed obliquely across it between the crural and internal abdominal rings. It then ascends, rather more vertically, between the latter and the triangle of Hesselbach (*d*). The obliterated hypogastric cord (*i*) passes obliquely forward and upward across the area of the last-named space, from the side of the bladder to the umbilical cicatrix. It is placed, ordinarily, to the inner side of the crural opening.

According to the calculations of Cloquet and Sir A. Cooper, in one out of every three instances, both in male and female subjects, there arises irregularly from the *epigastric* or from the

FIG. 58.

parent trunk near it, the *obturator* artery, which then descends behind and across the superior ramus of the pubis, to occupy its usual position in the sub-pubic notch. This irregular course is usually taken on the outer side of the crural ring, close upon the external iliac vein. (Fig. 58.) The precautions which are taken to avoid injury to the latter in operating will, under these circumstances, place the artery out of danger.

But sometimes its origin from the epigastric is placed at some distance from the parent trunk of the external iliac. The abnormal obturator then occupies a position to the inner side of the crural ring, lying in the sub-serous fascia behind the conjoined tendon and Poupart's ligament. When so placed in reference to a strangulated femoral rupture, it is exposed to great danger of being divided, if a free cut be made to the inner or anterior side of the seat of strangulation. In these cases, the irregular artery is placed usually at a greater distance from Poupart's ligament, in front of and above the hernial sac, than it is from Gimbernat's ligament to the inner side of the sac. This is owing to the length and upward direction of the epigastric artery before it gives off the irregular obturator, as seen in Figure 59. In this way, as pointed out by Sir Astley Cooper, more room is given for an incision in operating upon Poupart's ligament than upon Gimbernat's, and,

Fig. 59.

in cases of strangulation, with less danger to the artery. I have found by many experiments upon the dead subject uninjected, that on account of the very lax and moveable nature of the connexions of the vessel in the loose areolar tissue in which it lies, it yields easily to the pressure of the end of a blunt-pointed hernia knife introduced from below, and so avoids injury; unless the instrument be passed more than half an inch beyond the crural ring, and a fair cut be administered upon the tissues rendered tense by distension.

It will be seen that the *epigastric* artery itself is not free from danger, if Poupart's ligament be freely divided; but that it may be easily avoided by attending to the precautions just mentioned.

The *circumflex iliac* artery, by its external position and the secure protection afforded to it by Poupart's ligament and the iliac fascia, is remote from danger.

In a case of crural hernia figured by Sir Astley Cooper in his valuable work before cited (*Plate* XIX.), the obliterated hypogastric artery passes upwards to the outer side of the neck of the sac; instead of being placed in its usual position to the inner side. In such a case the sac is derived entirely from the false ligament of the bladder, and this viscus is not unlikely to be found partly within the hernial sac, if a large one.

XV.

CAUSES AND PATHOLOGY OF CRURAL HERNIA.

If a female subject be examined and compared with the male, it will be found that a considerably greater distance intervenes between Poupart's ligament and the margin of the bony pelvis opposed to it. In the male, this interval is usually completely occupied and securely packed up by the iliac and psoas muscles, and the vessels and nerves which pass with them into the thigh. It is, moreover, well protected by the junction of the strong iliac and transversalis fasciæ. In a subject advanced in age, the shrinking of these muscles, and the weakening of the crural arch and fasciæ which takes place, predispose more to the formation of crural rupture. Hence the greater frequency of this kind of hernia in old persons of both sexes. The greater space in the female results from the greater stretch and expansion of the iliac wings, increasing the distance between the iliac and pubic spines which afford attachment to Poupart's ligament. By this means, not only is the depth from the ligament to the pelvic brim increased, but to a still greater extent is the width of the opening amplified, to be filled up by soft parts. Since these do not usually present so much bulk as in the male, they are less effective as a stoppage in the larger opening.

It may also be seen that when the thighs are extended and abducted, the effect of the traction thus made by the fascia lata upon Poupart's ligament, is to increase its inverted curve, and to approximate it to the pelvic brim; thereby deepening the hollow of the groin and diminishing the diameter of the crural passage.

The aspect of the crural ring being almost directly upwards and downwards, the weight of the abdominal viscera and the pressure of the muscles upon them, have a simultaneous effect upon the peritoneum and the *septum crurale* covering the ring above. When, therefore, the ring and canal are wide, as in the female, and the bulk of the psoas and iliacus when contracting less effective in diminishing this width, a rupture of the crural kind is more apt to form. In males affected by crural rupture, it will usually be found that the pelvic wings have much of the wide, outstretched appearance of the female, and that the muscular system is less developed than usual. The tendency to crural rupture is increased after great distension of the abdomen by pregnancy, dropsy, tumours, or accumulation of peritoneal fat. This attenuates the structures which bind together the fibres of the tendinous openings, and enlarges all the existing apertures. In males, such distension acts chiefly upon the abdominal rings, because of their greater patulousness from the presence of the spermatic cord; and it diminishes the tendency to crural protrusion by pressure upon Poupart's ligament from above. Cases are, however, on record, and I have myself met with a specimen, in which both an inguinal and a crural protrusion had occurred on the same side.

In many cases, dilatation of the crural canal results, in the first place, from the growth of fat in the septum crurale and among the lymphatics in the canal. In the dissecting rooms it is common to remark, in fat female subjects, an atrophied condition of the fibrous and areolar structures in this region, disintegrated by the deposit of adipose masses, which are found projecting into and dilating the crural canal. These may even be mistaken for an omental hernia, from a certain impulse communicated to them by their passage from within the abdominal walls. A certain degree of fulness is communicated to the hollow of the groin by the presence of this fat in the canal, raising to a certain extent the enfeebled crural arch. This condition is usually accompanied by a facility or looseness of sliding of the layers of fascia over each other, which has been previously pointed out as favourable to the production of herniæ.

The first occurrence of the rupture is due to the yielding of the peritoneum, and to the loosening of its connexions through the sub-serous tissue with the upper margin of the crural sheath. It then becomes readily formed into a pouch, at the expense of the neighbouring folds or false ligaments of the bladder. The force of the abdominal muscles is then brought to bear upon the thin *septum crurale*. Having, in the female, a larger space to cover than in the male, and weakened perhaps by the interstitial deposit of fat in its meshes, it speedily gives way before the protrusion. The combined resistance of the structures being thus overcome, and the crural canal fairly encroached upon, its sides formed by the funnel-shaped sheath then give way more or less readily, and finally the cribriform fascia also yields, both affording a covering spread over the sac. The saphenous opening is then dilated, permitting the fundus of the tumour to emerge under the integuments of the thigh.

If the last-named tissues be more resistant, the rupture may remain for some time like a bubonocele in the canal, forming an obscure tumour behind the crural arch, in the hollow between the pubic spine and the femoral vessels. In this position it may dilate considerably the crural ring, and encroach upon the front of the femoral vein externally, by dilating the longitudinal septum. (See Fig. 60, *b*.) The direction taken by the rupture is first downwards, with a slight inclination forwards. Sometimes it passes downward to the full extent of the funnel-shaped sheath, as far as the lower border of the saphenous opening, to the union of the saphena with the femoral vein, without passing the cribriform fascia. In such cases, the latter structure is unusually tough and resisting.

Instances are recorded by Cooper and Hesselbach of a portion of the sac protruding through the dilated lymphatic apertures of this fascia, forming one or more offsets from the front of the sac-cavity, over which the coverings were thinner than in other parts, sometimes producing the effect of an hour-glass constriction. On emerging through the saphenous opening, the rupture alters its direction, and passes upwards and outwards towards the iliac spine, bending over the falciform margin of the opening, so as to lie over the femoral vessels upon the iliac portion of the fascia lata.

Fig. 60.

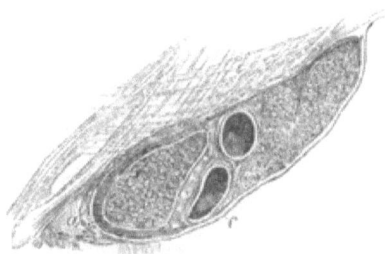

Occasionally, it may reach as high a position as Poupart's ligament, or even, by bulging before it the deeper layers of superficial fascia, expanding these structures over its distended fundus, it may reach beyond that line.

The cause of this change in direction is to be found in the very firm connexion which exists between the crural sheath and cribriform fascia, and the coats investing the great saphena vein and the chain of lymphatics at the lower part of the saphenous opening, which the first-mentioned structures furnish with a distinct sheath. It is partly also, doubtless, owing to the direction of the vessels and lymphatics which pass between the abdominal wall and the inner part of the saphenous opening, and which are bound together by a strong investment from the cribriform and deep layer of superficial fasciæ. In order to make its way down the thigh along the course of the saphena vein, the rupture must split the coats of this vessel, which are usually too closely combined to allow of this. In one instance, however, I have seen this condition reach to a considerable extent down the thigh. On the outer side of the saphenous opening, the laxity of the attachment of the different layers of fascia offers the least resistance to the protrusion, which, consequently, usually makes its way in this direction.

While the rupture remains moderate in size, the cylindrical outline and relations of the crural canal are in some measure preserved and impressed upon the neck of the tumour; while the fundus, after emerging through the saphenous opening, dilates into a globular tumour lying external to it, the whole presenting a bent or curved appearance (as seen in Fig. 54, B, page 192, in which the upper dotted line (*a*) represents the position of the crural ring, and the lower (*b*) that of the saphenous opening).

The pathological and surgical consequences of this arrangement are very important, and arise from the linear pressure made at the upper and outer parts of the bend of the sac upon the protruded bowel, by the edge of the falciform process of Burns' and Hey's ligament. This pressure sometimes forms an evident contraction of the sac of an old crural rupture at this part, exactly analogous to that before described as occurring in inguinal rupture by an undilatable and narrow superficial abdominal ring. By a misdirected pressure of the taxis ignorantly applied, or by extending and abducting forcibly the thigh of the patient when the hernia is strangulated, this pressure may be increased to a dangerous extent. If, by a misapprehension of the course taken by the tumour, a large crural rupture be pushed forcibly upwards towards the abdomen, the bowel may be injured by the pressure against the tense falciform margin. A similar effect may be produced by stretching the fascia lata in extension and abduction of the limb. By such means the femoral ligament, already stretched tensely across the sac, is made to indent the surface of its neck with such force as to risk the bursting of the included bowel. During a proper application of the taxis in a rupture of this kind, the fundus is first brought inwards, and a little downwards. It should then be compressed firmly, and pushed backwards through the saphenous opening. When it begins to give way, it will pass almost directly upwards through the crural ring. The principal force required in these movements will generally be the backward pressure, to cause the fundus of the tumour to pass the saphenous opening. At the same time, the fingers of the other hand are made to knead gently the neck of the sac at the crural opening, through which that part of the bowel which descended last will return first into the abdomen.

The ordinary coverings of a crural hernia, commencing at the skin, and taken in the order met with in operating, are as follows—viz., 1st. The *superficial fascia*, arranged at this part indistinctly in two layers, which together form generally the chief thickness of the hernial investments, and contain and support many small vessels, lymphatics, and nerves. 2nd. The *cribriform fascia*, known by its fibrous network, oval perforations, open texture, and its attachments to the margins of the saphenous opening. 3rd. The *fascia propria*, formed by the internal or lymphatic compartment of the infundibular sheath of the femoral vessels. This structure is the most dense and of the closest texture of any of the rupture coverings. 4th. The *septum crurale*, which is so thinned in hernia that it may practically be taken as a portion of the sub-serous areolar tissue; indicating, by the usual presence of a gland, and by the yellow colour and loose pellety texture of its fat, its close proximity to the surface of the serous sac of peritoneum immediately investing the viscera.

In a large rupture the deeper coverings may be more or less blended together, the superficial fascia being usually the most distinct. After great and long-continued truss-pressure, however, this also may be found blended with the deeper structures. When crural rupture affects the male subject, the course of the spermatic cord is obliquely across the front part of the neck of the sac, lying above Poupart's ligament, and crossing it just outside the pubic spine, and immediately over the inner part of the crural opening. On the skilful application of a proper truss, the cord escapes pressure by sliding to the inner side of the compressing pad. In operating upon a crural rupture in the male, this position of the cord must be borne in mind.

In an old standing crural rupture of considerable size, Poupart's ligament and the deep crural arch are usually found in a lax and enfeebled condition. The inverted convexity of the former is more or less obliterated, and the hollow of the groin is bulgy internally. The fascia lata is feeble and loose, permitting a freer and more unbalanced play to the muscles of the thigh. The upward and forward traction of the abdominal muscles, acting against the weakened fascia lata, tends still more to diminish the hollow curve of the groin, to enlarge the crural opening, and to alter the direction of its aspect by throwing its axis more forwards than normal. Thus the rupture is turned more directly to the surface, and the length of the crural passage anteriorly diminished. In extreme cases, occurring in fat subjects, this is so much the case that it becomes often very difficult to determine whether a rupture, when strangulated, be of the crural or inguinal variety, the tenuity and laxity of Poupart's ligament rendering it difficult to determine its position with reference to the neck of the sac.

CAUSES AND PATHOLOGY. 207

Another important change of relations occurs under these circumstances. The vein, instead of occupying its usual position on the outer side of the sac, becomes placed, by the encroachment of the latter outwards between it and the crural arch, almost entirely posterior to the sac, as seen in Fig. 60 (c), (page 203). It is very essential to bear this in mind in treating large ruptures by operation.

In a female subject met with in the dissecting-room of

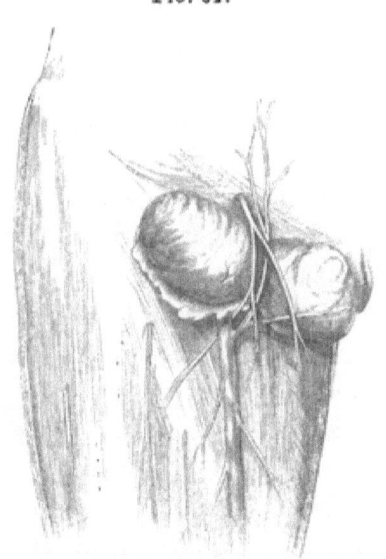

FIG. 61.

King's College, I found an irreducible femoral rupture constricted vertically across the middle, in such a manner as to render the long diameter of the sac oblique, strikingly resembling an oblique inguinal rupture. (Fig. 61.) The inner end of the tumour was placed close to the labium, encroaching to some extent upon the pudendum, and obscured by its fat. Upon dissection, the constriction was found to be owing to an indentation made by the ilio-inguinal nerve and superficial epigastric vessels, enclosed and supported by a tough band of fascia crossing the sac. The cavity of the sac was shut off from the peritoneum by adhesions at the site of the crural ring, and formed a distinct cyst. A case of complete separation of the sac from the peritoneal cavity, and its conversion into a cyst, is also recorded by Mr. Heath in the *Pathological Transactions* for 1861.

The occasional occurrence of such a termination to crural rupture, though less common than in the inguinal variety, gives a reasonable ground for believing in the possibility of producing such a result by proper pressure or by operation. Such an event is evidently more likely to occur in those cases in which the tubular shape and relations of the neck of the sac before described have not been entirely obliterated. As in oblique inguinal hernia, we have in these circumstances the advantage of a valve action of the anterior upon the posterior wall, and a more extensive approximation of the surfaces of the sac, upon which the effect of pressure or ligature is more likely to obtain a permanent resistance to redilatation. The retention of this tubular form seems to depend entirely upon the depth, strength, and resistance of that process of the fascia lata described under the name of the femoral or Hey's ligament. When this obtains a broad attachment to Gimbernat's ligament, and, as is sometimes the case, to a considerable extent of the surface of the pubic fascia lata below it, it may resist powerfully any extensive dilatation of the saphenous opening upwards, and may thus obscure the neck and throw the fundus of the hernial tumour more outwards. Occasionally it thus becomes the seat of a dangerous strangulation. This arrangement is usually associated with a low division of the common femoral vessels, and a low junction of the saphena with the femoral vein.

It is in such cases, more frequently found in the male

subject, that the hopes of the patient for a radical cure by the use of a proper truss may be rightly encouraged. On the other hand, the wider and shorter is the neck of the sac, and the more feeble and lax the structures about it, the more forwards is directed the axis of the crural opening, and the less is the chance of a radical cure being by this means effected. These conditions are, according to my experience, more commonly present in female patients affected with the complaint, in whom the hernial tumour is more evident, more moveable, and more distinctly isolated from the surrounding structures. Again, in the smaller and more recent cases, the sac of this variety of hernia is usually but loosely attached to the sides of the canal, and may be readily returned into the abdominal cavity after its contents are pressed up. A more complete adhesion of the opposing walls of the fascial crural sheath may then be obtained, either by pressure only, or, in more intractable cases, by the skilful application of a ligature in the manner presently to be described.

XVI.

DIAGNOSIS OF CRURAL HERNIA.

One of the most necessary distinctions to be recognised by the surgeon, either in operating for the radical cure, or for strangulation, or following a palliative treatment, is that between crural and inguinal hernia.

The sex of the patient may give a general idea, though many cases of inguinal hernia occur in the female, while crural rupture is not uncommonly seen in the male. It is more frequent in women who have borne children, and in advanced age, than in the young and barren. An unusual spread of the iliac wings in a male patient would indicate a greater probability of the crural form.

In thin persons a diagnosis is readily made from the low position of the rupture. In such cases Poupart's ligament is readily distinguished, and generally the principal fold or groove of the groin is plainly placed above the tumour. The crural canal and saphenous opening are felt distended and occupied by the sac; and, when large, its fundus or more bulky portion lies external to, and sometimes even above the neck; the very reverse being the case in inguinal hernia.

The shape of a crural rupture, as presented to the fingers of the surgeon, is usually more or less globular; that being the outline of the fundus, the rounded extremity of which is fairly presented, and most evidently perceived at the surface. (Fig. 62.) The shape of an inguinal hernia is generally oblong, the profile of the fundus and neck being both apparent at the

DIAGNOSIS.

surface. (See Fig. 4, page 36.) If large, the crural hernia assumes an oval shape, the long diameter of which is more or less transverse, while that of the pyriform inguinal is oblique, or more vertical than horizontal. This distinction is less apparent between a crural and a small inguinal hernia of the *direct* variety; but it is rendered the less necessary by the greater means of distinction, arising from the more internal and supe-

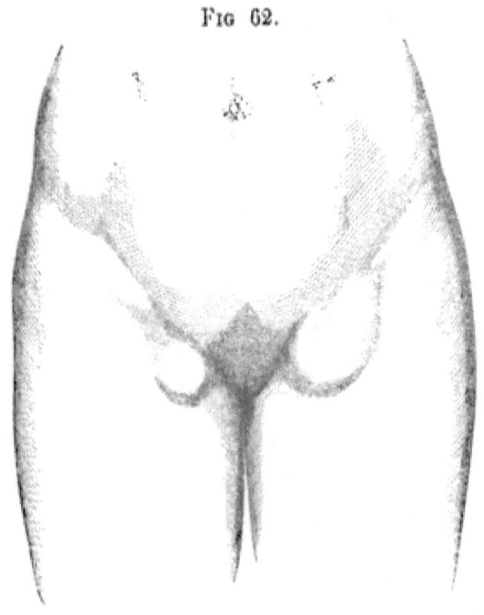

Fig 62.

rior position of the latter in reference to the pubic spine. In Figure 62, the crural rupture on the right side is small, and is just protruding the cribriform fascia; while, on the left, it is represented after having turned upwards and outwards.

In fat persons, these points are not so easily rendered certain; and it has doubtless happened in the experience of most surgeons that in cases of strangulation requiring an operation, the distinction has not been made clear until

Poupart's ligament has been exposed by incision. The position of the fold in the groin in such subjects is so uncertain in its relation to crural rupture as to be of no value in diagnosing doubtful cases. It may be depressed in position by the weight and volume of the overhanging abdominal fat, causing the attachment of the fascia to Poupart's ligament to sway, or belly downwards. The same displacement may be caused by a chronic enlargement of the upper chain of inguinal glands. The spine of the pubis affords the most valuable guide to the discernment of this point. This projection can usually be felt sufficiently plainly even in fat subjects. It lies to the inner side, and often a little above a crural rupture; and to the outer, lower, and back part of an inguinal. The anterior superior iliac spine can also, in all cases, be distinguished, being indicated in the fattest subject by a dimple or depression. If a line be drawn between these two points, and the bulk of the tumour lies below it, the case is, probably, one of crural rupture; if above, of inguinal.

The direction of the cough impulse, when present, affords some guide. In an inguinal hernia it is invariably greatest in a direction downwards, forwards, and *inwards*. In a crural hernia it is greatest directly downwards and forwards in the smaller cases; in larger, distinctly *outwards*; and in some it is even directed slightly *upwards* at the fundus.

In the male, the spermatic cord lies on the inner side, and rather superficial or anterior to a crural rupture; its position in the inguinal variety being precisely opposite.

A crural hernia, although lying upon or partly, in some large cases, even above Poupart's ligament, may be depressed at the fundus so as to leave a chink or fissure, admitting the fingers more or less, so as to distinguish this ligament above the tumour. In inguinal hernia this movement cannot be accomplished more on one side of the tumour than on the other.

In a crural rupture which is not readily reducible, much difficulty and pain is usually experienced in stretching, and abducting at the same time, the limb. This is sometimes experienced also in the inguinal variety when strangulated, but to a less degree. The pressure made by a crural rupture upon the femoral vein will, in many cases, produce an evident distension of the saphena during the abovementioned posture of the thigh; and when large and omental, a sense of numbness, or even œdema of the limb, may be to some extent present.

If the superficial abdominal ring can be felt by the finger above the pubic spine unoccupied by the tumour, while the saphenous opening is obscured by it, all doubt as to its crural nature will be at once removed.

The general signs of a hernial tumour, described in the section devoted to the inguinal variety as a means of diagnosis from tumours of a different nature, may be referred to in distinguishing the present variety.

A common source of doubt arises from the presence of *enlarged glands* in the crural ring and canal. If an enlargement affects the gland which lies in the *septum crurale*, especially if associated with enlargement of those placed above that structure, a certain degree of cough impulse may be experienced. If a similar condition is present in those glands placed within or below the saphenous opening, the double or multiple outline of the swelling is an evident means of distinction. The mobility of a chronic glandular tumour over the subjacent parts has been adduced as a means of detecting its nature. But if, as is often the case, the lymphatic ducts have themselves acquired, from long continued irritation, a certain degree of unyielding hardness and hypertrophy, and some of the glands within the cribriform fascia participate in the change, the mobility of the more superficial parts of the mass is diminished, or altogether prevented, and a sensation is given to the fingers like the neck of a sac containing hardened omentum emerging from the saphenous opening. The absence of the general signs of hernia; the history of the case; the state of the patient's health; or may be the presence of some source of irritation in the limb below, or in the penis and scrotum; or a similar glandular enlargement in some other part of the body, will lead usually to a proper diagnosis in such cases.

But it must not be forgotten that a small crural rupture may be co-existent with such a condition of the more superficial glands, and, when associated with symptoms of obstruction or strangulation, may require an exploratory incision to render a diagnosis certain, upon which the patient's life may depend.

Cases of *acute inflammation* of these glands in the first stage of gonorrhœa may occasionally give rise to symptoms somewhat resembling those of strangulation, such as constipation, nausea, and even vomiting. In doubtful cases, therefore, it is necessary to make a careful inspection of the genital organs, as well as of the whole leg of the side affected, to clear up the case, or narrow the area of diagnosis.

The elastic resilience and hollow percussion-sound afforded by a small sac containing bowel, is readily distinguished from the hard inelastic feel and dull sound of an enlarged gland. The sac is usually more globular, and the gland more elongated. But in an omental hernia, the contents of which are hardened by inflammatory deposit, there are more sources of doubt. The occasional presence of peritoneal concretions in a hernial sac has been before alluded to.

A *lobule* or *tumour of fat*, situated in the crural canal, and dilating by its growth the saphenous opening, is a condition occasionally found, and may closely resemble an omental hernia. The principal connexions of such a tumour are sometimes with the sub-serous fat of the peritoneum. As in the inguinal, so in the crural canal, such cases may be to a certain degree impressed with a cough impulse. Though not constituting or requiring the treatment of a rupture, such a condition may be followed by a hernial protrusion through the passages so dilated. I have been in the habit of recommending the application of a truss in such cases, not only as a preventive of hernia, but as a means of causing the absorption of the fat, by pressure constantly applied and securely fixed.

Most anatomists will recognise the occasional formation of hypertrophied lobules of fat in the meshes of the cribriform fascia. Unless such growths are large, so as to encroach upwards upon the crural ring, their diagnosis from hernia is not difficult. There is no cough impulse, and no fulness at Poupart's ligament affecting the outline of the fold of the groin. If large, a strict attention to the mode and history of their formation, and a careful examination of the crural arch, are necessary. Like other fatty tumours, they are much more moveable upon the imbedding tissues than a hernial tumour. If small, the pressure of a truss-pad as a precautionary measure can do no harm, and may result in the absorption of the mass.

Cysts occasionally form in connexion with the lymphatics of the crural canal. Occasionally, as before mentioned, these result from the occlusion of the neck of a small hernial sac. Their unvarying growth, rotundity, tenseness, mobility, and elasticity, and the fluctuation which may occasionally be felt, will usually suffice, with the absence of any implication of the crural ring, and of other signs of hernia, for a clear distinction. Such a distinction is, however, rendered the more necessary, that the application of a truss to this kind of tumour may do positive harm in opening up, by an effective application of fluid pressure, the crural canal and ring.

Chronic abscesses, glandular in origin or connected with disease in the lumbar region, pelvis, or hip-joint, may give rise, in certain cases, to some hesitation, by occupying the saphenous opening or crural ring. In ordinary cases of psoas abscess, the distinction of a more external position, often clearly on the outer side of the femoral vessels, is evident enough. In some cases, however, the matter finds its way on the inside of the muscle, behind the vessels, to the position of the lymphatic compartment of the crural sheath; or, by giving rise to suppuration in the lymphatics, may occupy the crural and saphenous openings, and result in a tumour having more or less moveable coverings, a distinct cough impulse, and some increased distension when the patient is in the erect posture. Pelvic abscess has also been known to point in this situation. In such cases the previous history, the effect on the patient's general health, the accompanying pains in the back, loins, or pelvis, the hectic and other symptoms, are usually marked enough to rouse the most unready suspicion. The subsidence of the tumour in the recumbent position, however long continued, is never considerable, the fluctuation is usually sooner or later made evident; while the enlargement is gradual, invariable, and, in the latter stages, more rapid than in hernia; of which also the other signs are absent. The concomitant association of hip-joint disease is usually sufficiently apparent, where that is the cause of the swelling.

XVII.

OPERATION FOR THE RADICAL CURE OF CRURAL HERNIA.

IN the *Lancet* for May 2nd, 1828, a case of operation for the radical cure of a crural hernia is recorded as performed by Dr. G. Jameson, of Baltimore, North America; in which, after an operation performed upon a lady for strangulation, leaving the rupture still liable to protrude, this surgeon, following apparently the operation proposed by Dzondi, dissected from the skin of the groin a lancet-shaped flap, which he then forced into the hernial aperture exposed by the incision, finally stitching over it the edges of the wound. From the concluding remarks on the case, its termination does not seem to have been so satisfactory to the patient as to the operator; whence one is led to infer that it supports the conclusions previously expressed, as to the inutility of dealing only or chiefly with the integuments covering a rupture.

This case is the only one I can find recorded of an operation undertaken for the radical cure of this kind of hernia, with the exception of one performed by Mr. Redfern Davies, of Birmingham, a modification of Wurtzer's operation, upon the principle of invagination of the skin into the crural opening, by a hard plug retained by sutures. The death of the patient, which followed some time after the operation, appears, from the account given of the case by this enterprising surgeon, not to have been directly connected with the proceeding.

An operation which I have been accustomed to show on the dead subject for the cure of this form of rupture, is performed as follows,—the instruments used being the same as for inguinal hernia by the wire operation.

The patient being placed in the same position as for the preceding operations, and the rupture completely reduced, a vertical incision, about an inch long, is made through the skin over the site of the tumour. (See Fig. 65.) In Figures 63 and 64, the integument is represented turned down by a

Fig. 63.

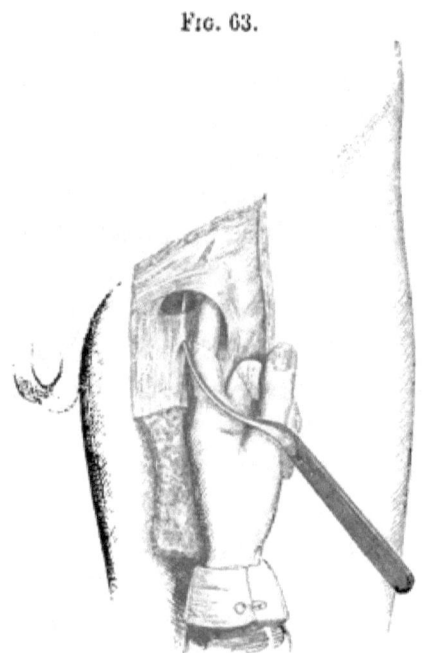

dissection, to show the exact position of the finger and instruments beneath it. The fascia around the margin of the wound is then detached in the usual way, to a sufficient extent to allow it to be invaginated, free from the skin, fairly into the crural opening.

The forefinger used for invaginating is then placed upon the inner side of the femoral vein, pressing it carefully outwards out of danger from the needle. This instrument is then passed backwards through the sac, sufficiently deeply to take up the pubic portion of the fascia lata covering the pectineus muscle; the point of the needle being afterwards made to appear in the wound. (Fig. 63.) It is then carried forwards and upwards, and made to transfix Poupart's ligament close to the nail of the invaginating finger. The skin of the groin is then drawn outwards by an assistant, and the point of the instrument pushed

Fig. 64.

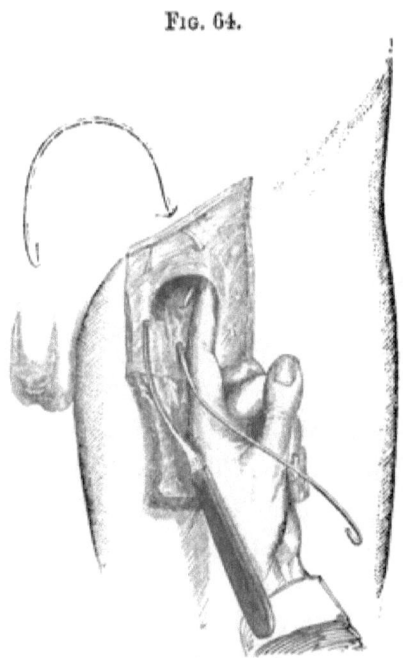

through it. The wire, prepared in the same manner as in the operation before described, with a hook at each end, is then hooked on to the eye of the needle, and drawn back with it into the wound. Then the needle, being disengaged from it, and leaving it protruding through the incision, is passed again through the pubic portion of fascia lata, about an inch or three-quarters of an inch (according to the size of the hernial opening) to the inner side of the first perforation (Fig. 64); and its point made in the same manner to pass through Poupart's ligament directly above, and close to the curved border of Gimbernat's ligament, of which a portion of the fibres may be included. The skin is then drawn inwards, until the needle can be pushed through the same puncture before made, which is already occupied at this stage of the proceeding by one end of the wire. The opposite end of the wire is then hooked on, and drawn back with the needle, which is then finally disengaged. The two ends of the wire are now twisted down into the incision, and cut off about six inches from the twist. The loop which emerges at the upper puncture is then twisted firmly down into it, pressing down before it the portion of Poupart's ligament opposite the crural opening, which is included in its grasp. (Fig. 65.)

In a small hernia the ends of the wire may be bent upwards over a pad of lint rolled tightly, and hooked into the loop above (B.) In a rupture of larger size, where the ligamentous structures are much relaxed, a more steady compression will be achieved by the use of a boxwood or glass compress of the cylindrical form, (depicted in the figure at A,) and described in the operations for inguinal hernia. The upper end of the pad is to be placed in the embrace of the upper loop; and over the lower end the extremities of the wire are fastened by a turn or two. In cases very large, two wires may be employed in the same way, and twisted over a double pad, or two pads, laid side by side, so as more effectually to occlude the nec of the sac.

Pledgets of lint placed on each side, and a spica bandage over all, complete the dressing.

By this operation, that part of the tendinous crural arch which overrides the neck of the sac, is drawn backwards and downwards, and becomes adherent to the pubic portion of fascia lata included in the suture. If the rupture be large, the sac and its coverings are transfixed, and embraced in the suture; if small and recent, the serous sac may be returned easily within the abdominal parietes, and its tendinous investments united altogether external to it. The invaginated fasciæ and sac coverings become adherent on all sides under the combined influence of ulceration, traction, and pressure; and thus form a consolidated mass, filling up the area of the crural ring.

The *precautions* which are chiefly necessary are—first, to

FIG. 65.

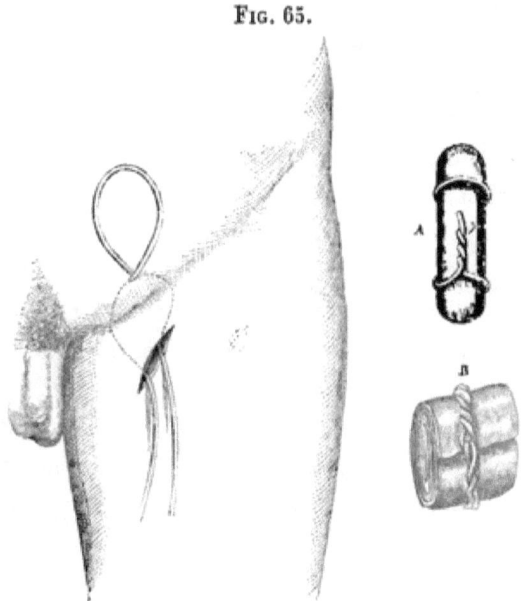

keep the finger carefully pressed upon the site of the femoral vein, so as to interpose between it and the needle during its passage; and—secondly, to avoid pushing the instrument too far into the abdominal cavity, by which the bowel, the epigastric artery, and in the male, the spermatic cord, might be endangered.

The wire may be kept in for the same time, and removed in the same manner as before described for inguinal hernia.

In severe cases of crural rupture affecting the male, or in healthy females whose laborious occupations may necessitate great muscular exertion, I believe the foregoing operation to be the most likely to accomplish a cure, or at least a diminution of the rupture, so as to render a truss available, with the smallest amount of risk and inconvenience to the patient of any that have hitherto been proposed.

The close proximity of the large femoral vein to the site of the operation, and the implication of the numerous lymphatics of the part, seem to afford a greater chance of bad results from the subsequent suppurative action, than is the case in inguinal hernia. The same risks, however, are frequently encountered without bad results, in operating for strangulation.

The operation seems to be more likely to succeed in effecting a complete cure, in the cases previously described as characterized by a considerable extent of the hernial canal, between the crural ring and the falciform margin of the saphenous opening. In such cases, the depth of the upper horn of the opening before described under the name of the femoral or Hey's ligament, affords a broader surface of apposition for adhesive action, and consequently a greater resistance to future protrusion. Speaking without experience upon this point, it may be said that, in the majority of cases of crural rupture, the shortness of the crural canal investing the neck of the sac, and the more direct downward pressure of the intestines brought to bear upon the area of the adhesions which may be obtained by operation, will render a cure more difficult to accomplish than in the inguinal variety. A diminution of the opening may, however, be thus obtained, if not a complete cure; and thus a more effective support be thenceforwards obtained by means of a proper truss.

It is, I think, a question which may be fairly entertained, whether, after an early operation for strangulated hernia of both the inguinal and crural varieties, where there is little expectation of inflammatory action in the bowel released, wire sutures may not be advantageously employed in drawing together the edges of the tendinous opening, denuded and severed by the incisions, and so accomplishing a radical cure by the same operation. This might be done with still less risk in cases where the operation is accomplished without opening the sac; since in these instances the small chance of mischief which may be expected to result from an implication of this offset of the peritoneum may be taken away by returning the sac within the abdominal parietes. This may usually be done with great ease in cases of crural rupture of moderate size and duration. It needs hardly be said that the advantage of obtaining a radical cure, which so rarely, if ever, follows from the operation for strangulated hernia as usually completed, is, in favourable conditions of the patient, amply worth so small an increase of risk as might ensue from symptoms so slight as those produced by the application of wire sutures in this way, in the great majority of the cases.

XVIII.

PROSPECTS OF A RADICAL CURE BY TRUSS PRESSURE IN CRURAL HERNIA.

If the occurrence of a complete cure by external pressure only is rare in inguinal hernia, such a result is still less frequent in the crural variety, for reasons which will be readily appreciated by a perusal of the foregoing pages.

That it does occasionally occur, however, is unquestionable. I have myself examined the bodies of two persons at least, in which the sac of a crural rupture had been cut off by adhesion from the cavity of the peritoneum, as before mentioned.

It seems to me that the way in which such pressure is applied by the ordinary form of truss pad is little likely to accomplish the desired result. In the truss as usually employed, the pad is applied obliquely from above downwards and inwards. A reference to the anatomy of the crural canal will show this not to be in accordance with the direction of the axis of the neck of the hernial sac, which is almost directly downwards, with a slight inclination *outwards*. The widest part of the pad is usually at its lower extremity, which presses into the saphenous opening; while the narrow part is but imperfectly applied to the upper part of the canal and crural ring, and, from the manner of its connexion with the side spring, has a constant tendency to slip away outwards from it.

It is essential to the production of a radical cure that the bowel should be entirely shut out of the sac cavity for a sufficient length of time to permit the sides of the latter to become adherent, or so contracted as not again to admit of protrusion. For the permanent duration of the cure, it is necessary, further, that the surrounding tendinous structures should also contract upon and close up the neck of the sac.

To accomplish this in a case of crural hernia it is essential to make the chief pressure upon Poupart's ligament, so as to close the crural ring by approximating the crural arch to the ramus of the pubis. The most prominent and widest part of the pad should be placed then in the groove of the groin, in order to shut out the bowel completely from the sac. To guard the saphenous opening is a secondary consideration, and one which will be necessarily carried out by the above means.

The objections to a convex truss pad, which were shown to be serious in the case of inguinal rupture, do not, in the present case, so strongly apply to such a pad when properly placed. The axis of the crural ring and canal coincide, and are directed upwards and downwards, and for practical purposes may be considered as vertical. (See Fig. 54, A, *a*.) That of the saphenous opening is directed forwards and slightly downwards and inwards (*b*). The pressure of a truss pad should fall chiefly exactly between the crural ring above and the saphenous opening below, and at right angles to the axis of the canal (*c*). Its effect would, in such a place, be to close the sides of the canal, and not to dilate it by working into it in the direction of its axis. If the pad, however, be improperly directed and placed too low, and exerts its chief and most convex pressure upon the saphenous opening, it will invert the soft parts into the opening, and tend to dilate it and work its way into the axis of the canal.

The truss pad I have had made for femoral hernia is composed of the segments of two ellipses (Fig. 66, A), joining a little above the centre of the pad at its broadest part (*a*). The segment of the upper ellipse is wider from side to side, and that

ON CRURAL HERNIA.

of the lower longer from above downwards, tapering off below, so as to accommodate itself to the shape of the hollow between the muscles of the part (*b*). In section (Fig. B) the pad is seen to be thicker towards the upper border at *a*, thinning off in a curve like the prow of a boat at *b*, while the central bearing point at the middle third (*c*), opposite to the insertion of the side spring, is flattened. The upper segment and the most convex and widest portion are placed over and upon Poupart's ligament in the groove of the groin, exerting pressure upon that structure. The lower part of the inferior segment

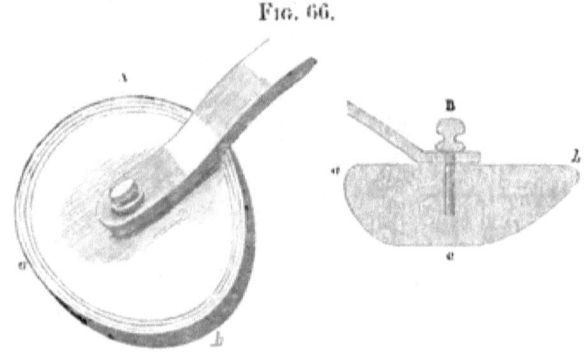

Fig. 66.

A & B. The Elliptical Pad and Section.

covers and protects the saphenous opening; while the flat central third rests on and compresses the femoral or Hey's ligament, which forms the anterior boundary of the crural canal.

For hospital purposes, and in ordinary cases, the pad thus shaped may be made of boxwood, and held in place by a spring passing round the hip of the same side.

In order to be able to regulate the pressure on any side of the pad especially, at which the rupture may tend to escape, or to increase its effect upon Poupart's ligament, the same arrangement as before described in the section on inguinal hernia may be employed. The basis of the pad is a thin plate of metal, shaped to the elliptical form, and raised at the borders into a cup-shape. A Y-shaped spring, like those previously described in the trusses for inguinal rupture, supports by its central lever the extremity of the side spring or belt, and is fixed to it by an adjusting screw. (Fig. 67, A and B.) The upper limb of the spring is connected by the hinge close to the margin of the upper ellipse (at *b* in the figure and section), pressing it firmly against the crural arch. The two lower press upon the longer ellipse covering the saphenous opening, and bearing evenly upon the sides, preserve a steady lateral

Fig. 67.

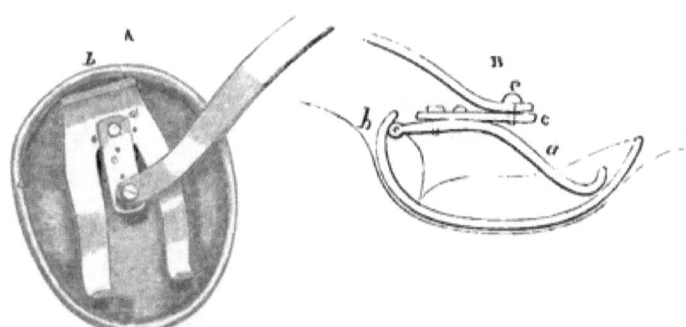

A & B. The Elliptical Lever-spring Pad and Section.

balance. The force of the hip spring can be directed more upon one or other side either by means of holes in the lever plate simply, or by a revolving slide and hand-screw, as represented in the figure. By this simple means we have a more steady and perfect control over the lateral motions of the pad than can be obtained by the ordinary ball-and-socket joint or spiral spring. The action of the pad spring, as seen in the section at B, has a constant tendency to adjust the pressure of the side or hip spring under the changing conditions of flexure of the body upon the thigh in sitting or standing, and to keep the pad always in the groove of the groin or deepest part of the surface to which it is applied. With a view to this effect, the hinge fixing the upper part of the Y-shaped spring is placed higher upon the margin than in the trusses for inguinal hernia. By reference to the figure (B) it will also be seen that the sides of the cup at the junction of the two ellipses are cut away, so as to allow of a slight adjusting motion, by bending the thin metal connecting the two together at the groove of the groin. This enables the maker to accommodate the bearing of the upper portion to the degree of protuberance of the abdomen of the patient; and to obtain, if necessary, a greater proportionate amount of pressure upon Poupart's ligament and the crural ring. A very steady bearing of the pad may thus be obtained, and any tendency to displacement upwards by the traction of the side spring be also, in great measure, counteracted.

In this form of rupture the action of a hip spring is generally even more required than in the inguinal variety, in order to keep the pad secure in all positions of the limb. When a simple body belt is used, it is necessary to combine with it a strap round the top of the thigh. This is most effectually accomplished by making the two cross the pad in a figure-of-8 way, with one upper and one lower end to fasten upon it. But in most cases this is irksome in the erect posture, and becomes relaxed in bent positions of the thigh.

XIX.

PREVENTION OF INGUINAL AND CRURAL HERNIA.

It has been found that inguinal ruptures are nearly three times more common than femoral, and six times more frequent in males than in females. It is especially important, therefore, for persons of the former sex, who are liable to be called on by the nature of their pursuits for sudden exertions and violent muscular efforts, to use such precautions as may diminish the tendency to the formation of rupture, and particularly to guard against inguinal protrusions.

Soldiers, sailors, mechanics, colliers, and labourers, and the numerous class of younger men who cultivate active and athletic sports, such as cricketing and boating, are peculiarly subject to this disagreeable consequence. This is especially the case if such sports be violently indulged in without a probationary and gradual exercise of the muscles, accustoming them and their associated tendons gradually to the demands of extraordinary exertion and pressure. Among our soldiers, the practice of riding with a military seat is justly considered to produce many cases of rupture. Hard riding of any kind, especially on heavy trotting horses, is to be avoided by those who have a predisposition to the disease. The hot, close atmosphere in which most of our mechanics and forgemen pursue their avocations, producing violent and continued perspirations, has a very relaxing effect upon the tissues, which makes itself evident in the pale and sallow countenance, and predisposes to hernial protrusion; while the violent muscular efforts usually required in their work supply the exciting cause.

The practice of training for boat races in the young men of our universities will sometimes produce, in individuals previously fat, an open condition of the abdominal rings by the sudden absorption of the fat which lodged therein, which is very likely to give egress to a rupture during the violent tension of the exciting contest. More than one instance of the kind has come under my own observation.

Hereditary predisposition to rupture ought always to be a significant warning. A certain conformation of the muscles and tendons and bones about and in the groin, descending to father and son, gives, as before pointed out in this volume, a strong predisposition to rupture.

One of the most important of the appearances which indicates a tendency to *inguinal hernia* is the fulness of the groin over the internal or deep ring, increased by coughing or holding the breath, and caused by deficiency in the development of the internal oblique muscle. Another, not less important, is a patulous condition of the superficial ring, which is often sufficiently open to admit the point of the finger previously to the appearance of a decided rupture. A lax condition of the scrotum and coverings of the testicle is often co-existent with such indications of predisposition. Varicocele and hydrocele, if neglected and suffered to grow very large, may prepare the way for a rupture by opening up the inguinal passage.

Crural hernia is to be apprehended in bodily conformations where the iliac wings are wide apart, increasing the length of Poupart's ligament; the hollow of the groin not well marked by a concave curvature, indicating a loose condition of the crural arch; and when the muscles of the thigh are loosely held together, indicating feebleness of the fascia lata. An open condition of the saphenous aperture, and a pulsation of the common femoral artery, characterized by a certain degree of mobility and distinctness, may be taken to show a patulous condition of the crural ring and passages. This may be increased also by a chronically enlarged condition of the femoral glands, which, on their subsidence, may leave space for the formation of a rupture.

A general increase of obesity, especially if affecting mainly the omentum and other contents of the abdomen, associated as this condition often is with a laxity of the fibrous structures, produces a degree of tension upon the abdominal walls which often terminates in this condition. The interstitial deposit of fat disintegrates more than anything the close texture of the fibrous aponeuroses of the body. At that period of life which is often characterized by corpulency, I have met with cases in which there has been a reappearance of a congenital hernia which had been considered as cured, and absent for a long period.

In the above cases the use of a light preventive apparatus over the groins during exercise of any kind is much to be recommended. An appropriate apparatus of this kind is made by Mr. Matthews, of Portugal-street, Lincoln's-inn-fields, after my own design.

It consists of an ordinary elastic suspensory bandage for the scrotum, which is buttoned on each side to a belt passing over the groins, and encircling the hips from one side to the other. Embodied in, or connected with the belt upon each groin is an elastic hollow pad, made of vulcanized india-rubber upon a metallic basis, shaped and adjusted so as to cover both the inguinal and crural rings. These may be inflated with air, or filled with water, so as to exercise a greater or less area of pressure, as the case may require. The pads are connected together and braced up by a strap and buckle passing from one to the other across the lower part of the abdomen above the pubis. (See Fig. 69.) A double perineal or under strap fastens on to the lower part of the pads on each side, and is secured to the back part of the body strap above each hip. In some cases it will be sufficient for these to fasten to the back part of the suspensory bandage. In all cases it will be better for the two straps to be connected to each other behind the scrotum before reaching the body belt. This apparatus may be worn without making any external show, and acts as a support to the parts liable to rupture, without weakening them by undue pressure, while its elasticity readily accommodates it to the different movements of the body and limbs.

Fig. 69.

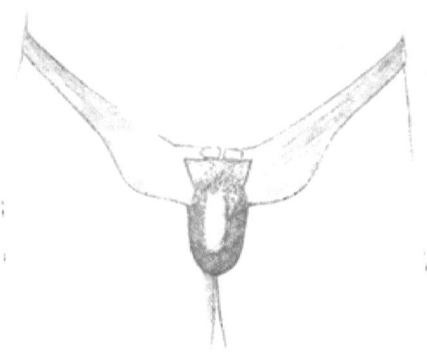

Patients with a predisposition to, or actually affected by, rupture should avoid a flatulent diet, and the habit of drinking large quantities of fluid of any kind. They should carefully look after the condition of the bowels, and avoid constipation by a due use of aperients or regulation of the diet. A tendency to the accumulation of fat should be checked by the use of a spare, dry diet, and walking exercise daily. The habitual use of hot or Turkish baths, and excessive smoking, should be avoided, as tending to a relaxation of the tissues. Cold plunge or douche baths, on the contrary, are very beneficial, having an especial effect upon the genital organs and the structures in their neighbourhood, as may be seen by their effect in bracing up the testes and scrotum by inducing vigorous contractions of the cremaster muscles and dartos. In lax conditions of the genitals and groin, with more or less varicocele, the administration of muriated tincture of iron will be found of great use, aided by the cold douche applied daily.

For young boys having a tendency to rupture, a moderate use of gymnastics of a kind to employ chiefly the arms and legs, is decidedly advantageous, as tending to strengthen by tension the muscular and ligamentous structures. All feats of strength, however, should be carefully eschewed, together with such tricks as balancing on the abdomen over a pole, or sustaining weights upon it, and all inordinate action of the abdominal muscles. Open air bathing and swimming are also excellent, but should be taken in moderation, and never to great fatigue.

For working men who have a tendency to rupture nothing can be better than a liberal use of the cold water douche daily all the year round, over the groins and genital organs. The tonic action of such cold affusion is speedy, palpable, and enduring, and conduces to health and strength, both local and general. The bowels should be carefully regulated, and all things tending to flatulent distensions avoided.

XX.

TREATMENT OF IRREDUCIBLE HERNIA.

As a rule, when a hernia is irreducible from adhesion or incarceration, no attempts should be made to effect a radical cure by pressure or operation; but if the tumour be of small size, and it be clear that the contained and adherent mass consists of omentum only, much good result may be expected from wearing a proper truss. In such a case the spring of the truss should not be too strong, and the pad should be flat on the compressing surface, so as to exercise little pressure upon the adherent omentum lying in the axis of the hernial canal. In addition to blocking out the bowel, such pressure may produce ultimately an extension of adhesive action all round the neck of the sac, so as completely to obliterate its communication with the cavity of the peritoneum. In these cases, decided invagination of the anterior coverings of the hernia under truss pressure is not so apt to take place, on account of the presence of the omentum filling the sac. If the contained mass be hardened already near the neck by consolidated effusion, hope may reasonably be entertained of an ultimate cure by this means. Great care must, however, be taken lest a portion of the bowel also slip into the sac, behind and obscured by the omentum, and thus become exposed to, and injured by the pressure of the truss. The occurrence of such an addition to the protrusion will generally be indicated by the accompaniment of abdominal disturbance, constipation, unusual pain, and sudden increase of size in the tumour, and inability to wear the truss comfortably. It has been thought that the pressure of a truss upon adherent omentum would be likely to give rise to unpleasant symptoms. I have, however, frequently observed patients having a portion of omentum within the sac able to bear the pressure of a truss upon the mass. The pressure must not be too strong, but must be regulated by the feelings of the patient. If too powerful a pressure be applied, a congestion of the omentum may be induced, which may give rise to symptoms. Generally speaking, the feelings of the patient are the best guide in this matter, and will effectually prevent so much or so long a pressure as will interrupt the circulation through the incarcerated portion of omentum. A moderate degree of pressure, on the other hand, will not only prevent the further accession of omentum or of bowel into the sac, but will tend to promote absorption of that already adherent, or to cause a more complete blocking up of the hernial canal.

The cases placed on record by Pott, Cooper, and others, in which interference with the omentum has given rise to symptoms of abdominal disturbance, have apparently been owing to a degree of traction exercised upon this structure sufficient to interfere with the free movements and functions of the stomach and colon, to which it is connected above. Such symptoms, however, can scarcely arise from a moderate degree of pressure upon an omentum long adherent to, or contained in, a hernial sac, without having previously exhibited these effects. If such symptoms have been occasionally induced by the irreducible rupture, the case must not be considered as one favourable for treatment by this means. There seems to be nothing in the structure and formation of the omentum itself which should prohibit some degree of pressure upon it, since its nervous supply is not plentiful. Numerous cases are recorded, and occur in the practice of most surgeons, in which a portion of the omentum has been strangulated by ligature, or in a state of mortification, or removed altogether, without producing symptoms. Opinion at the present day appears to be recurring to that entertained in the times prior to the greater authorities on this subject—viz., that the omentum, if properly dealt with, is much more amenable to interference than was inferred by them on the authority of some peculiar cases. On the other hand, the number and arrangement of its vessels, and the nature of its usual chief constituent, fat, would indicate its capability of bearing considerable pressure, provided it does not strangulate, and also its facilities for becoming adherent to the opposing surfaces of the sac without interruption to its vascular supply.

In certain favourable instances in which the adherent omentum, together with the sac itself, can be pushed up into the internal ring, the judicious application of a wire ligature below and outside the sac, embracing and closing up the tendinous walls of the canal, may produce a closure of the passage, which will prevent the descent of a further protrusion of the abdominal contents. The adherent omentum and sac will then block up the internal hernial opening and canal. Such a case will be found in *Case* 60 (*Appendix*). Several cases have come under my observation, both in the living and dead subjects, in whom an adhesion of omentum at the deep ring has effected a complete occlusion, and produced a radical cure. (See page 143.)

When, however, a portion of bowel is permanently retained in the hernial sac, or the retention of such omentum as habitually resides there gives rise, under certain distended conditions of the stomach and colon, to unpleasant symptoms, it is evident that nothing should be attempted beyond the support of the rupture by a bag, or properly fitting hollow truss-pad. These are best made of some elastic material, such as vulcanized india-rubber, of such a strength as to resist the action of the abdominal muscles without causing distress to the patient. This should be held on by straps round the hips, upon the general plan of the arrangement before described in connexion with the preventive belt and pads, with such additional perineal and thigh straps as the nature of the case may require.

The precautions of diet and constitutional treatment, before sketched out as necessary to be observed in hernial conditions, are especially to be regarded under the circumstances of irreducible rupture.

PART III.

UMBILICAL HERNIA.

XXI.

CAUSES AND PATHOLOGY.

So common is this form of protrusion, that Sir Astley Cooper considered, that in point of frequency it comes after the inguinal variety. It is certain, that including the infantile form, it is more commonly seen by the surgeon than crural rupture.

In the adult it is not so common, and usually occurs as a consequence of pregnancy, ascites, or a large accumulation of fat in the different processes of the peritoneum, especially the omentum, which is placed directly behind the navel. The distension resulting from the corpulency thus produced is more evident in the central than in the lower parts of the abdomen, and distends laterally the transverse abdominal section in which the navel is placed to such an extent as to stretch the fibrous structures, and to open up the normally small apertures which there exist. The weaker parts naturally give way the soonest, and admit of the protrusion of the abdominal contents. In most cases, probably a somewhat imperfect closure of the umbilical aperture in early infancy has preceded, leaving a weakness of resistance at this part. This is perhaps, in the adult class of cases, increased by the interstitial deposit of fat in the fibrous parts around the cicatrix of the navel itself, which contributes to diminish its resisting power.

In the infant it is very frequent, and usually occurs from the second to the fourth month after birth, while the aperture for the umbilical cord is still open. It is often in consequence of ill health retarding the tendency of the aperture to close up in regular progress of development, and at the same time giving rise to such suffering as to keep up frequent and violent fits of crying, during which the abdominal muscles exercise spasmodic pressure upon their contents, and force them continually into the aperture, thus keeping up a constantly recurring dilatation.

In the development of the fœtus the last portion of the peritoneum to close into a shut sac around the abdominal viscera is the central part, at the navel. At an early period in the development of the ovum an opening in this part of the abdominal walls transmits the *omphalo-enteric duct*, or pedicle connecting the *umbilical vesicle* or *yolk sac* with the intestinal tube; the *omphalo-mesenteric vessels*, which spread over it; together with the *urachus*, or tube of communication between the bladder and *allantois sac*; and the accompanying *placental vessels*. Of these structures the two first mentioned pass through the cavity of the peritoneum, to be connected with the intestines and their mesenteric vessels respectively; and the two latter pass outside and in front of the peritoneal serous membrane, the umbilical or placental vein upwards to enter the fold of the falciform ligament of the liver, finally to join the portal and inferior caval veins; and the placental arteries accompanying the urachus downwards to the bladder, on the sides of which they are continuous with the hypogastric branches of the internal iliac arteries.

After the third month of intra-uterine life, at periods varying somewhat with the vigour of the development, the duct of the umbilical vesicle and its associated vessels become atrophic, and disappear. The peritoneal sac is then entirely closed up at the navel, where a fine cicatrix may, at an early stage, be afterwards observed. The aponeurotic tendons which invest this part do not thereupon also entirely close, but leave a considerable aperture up to the time of birth for the transmission of the umbilical arteries and veins to and from the cord.

After the latter is tied and cut at birth, the remaining stump usually dries up completely to its root of attachment to the abdominal parietes, and its vessels are obliterated; the fibro-areolar remains of these structures completing the closure of the opening by forming a very tough and resisting connexion closely adherent to the peritoneum, the margins of the tendinous opening, and the neighbouring skin. The opening in the tendon at the linea alba is then normally reduced to the size of a small quill; and the tissue covering and surrounding it is very tough and difficult to be cut or pierced by a needle, even in very young children; affording a hold to a suture very valuable in any endeavour to close, by this means, an abnormal dilatation of the aperture. The close attachment of the skin to the tendon around the margins of the opening, together with the absence of subcutaneous fat at the part, give rise to the peculiar indentation of the umbilical cicatrix.

When the umbilical cord is large and tumid, its vein larger than common, and a slowness of development usually associated with this condition is coexistent, the opening in the tendon is larger than normal, the skin is loosely adherent, the child generally imperfectly developed, and a protrusion may take place even during the first struggles of the infant, and the bowel thus may be endangered by the ligature placed upon the umbilical cord. In these cases of early formation, the external structure of cord itself is distended near its junction with the skin of the abdomen, and forms part of the investment of the tumour before it has had time to shrivel and dry up.

Generally, however, the protrusion takes place at a later period, and the cicatrix formed by the dropping of the navel-string is found upon or near the centre of the hernial coverings. As the tumour grows larger it is increased at the expense of the surrounding integuments, by the yielding of their attachments to the margins of the tendinous opening.

When formed at an early period, the peritoneal investment or sac of the rupture is necessarily very thin and attenuated, produced as it is by the dilatation of the most recently formed portion of the serous membrane. Still it may usually be distinctly demonstrated and separated from the superjacent layers, upon which it is in some degree moveable. This sac, in almost all cases, contains omentum, which completely covers the bowel in this situation. In young children, this structure is very thin, and contains but little fat, and the contents of the hernial sac are chiefly made up of a portion of the transverse colon or small intestine. The usual presence of the former is due to its position exactly across this part of the abdominal cavity.

When formed at an adult period of life, umbilical rupture is most frequent in elderly women who have borne children. In females of a conformation producing a great anterior projection of the abdomen during pregnancy, such as a limited expansion of the iliac wings of the pelvis, or a great anterior curvature of the lumbar spine, often seen in women of small stature, an umbilical rupture is more apt to occur. The carrying of twins, or multiples, or of large children, or dropsy of the amnion, may also produce it, especially if negligence in bandaging be displayed. Generally it is due to a combination of the distension of pregnancy with that of an inordinate accumulation of peritoneal fat. Occasionally it is present from the latter cause associated with a hypertrophied condition of some of the viscera, or an habitually distended condition of the stomach in corpulent men, who are fond of the pleasures of the table. Sometimes it is due to dropsy of the peritoneal cavity.

By such an extreme distension of the abdominal walls the cicatrix closing up the umbilical opening is gradually thinned and weakened, until it yields before the pressure of the bowels. More frequently, according to Scarpa, the opening in adults occurs not exactly in the site of the umbilical cicatrix, but a little to one side of it, through an aperture in the tendinous septum of the sheaths of the recti, by a combination, doubtless, of distension from within and a deposit of fat between the fibres, in the course of the small vessels and nerves which traverse the part. At first, and in the smaller cases, a sac of serous membrane is distinct, and easily demonstrated. In some very large cases, however, it has been said by Cooper, Lawrence, and others, to be incapable of demonstration over the whole inner surface of the tumour, having become attenuated to a degree, and absorbed or burst through by the pressure. This result, occurring so much more commonly than in the other forms of rupture, appears to be due to the more extensive and intimate adhesion of the peritoneal membrane, and the greater thinness of its investment on the inner surface of the abdominal wall at this part than in the groins. At this part the subserous areolar tissue and fat is usually entirely wanting, and the membrane has but little scope of motion over the surface of the parietes; consequently, the peritoneum available for a sac is scanty, and becomes much attenuated by stretching, until it finally disappears by a process somewhat similar to that which ensues to the inner and middle coats of an artery subject to aneurism. A case is recorded and figured by Sir A. Cooper, in which the omentum passed in two separate places, through openings produced in the sac by attenuation or rupture. (*Op. cit.*)

At this period of life the contents of the sac are often a mass of fatty omentum. In large cases, however, it usually contains in addition a portion of colon or small intestine. The omentum is very apt to become adherent from inflammation set up by the pressure and friction of the clothing, or occasional violence, to which the position of the rupture renders it very liable.

When the rupture is large, the recti muscles become separated from each other at the linea alba to a considerable extent. In the normal condition, the borders of these muscles at the umbilicus are somewhat separated by the interposition of the umbilical cicatrix, the separation being continued upwards to the ensiform cartilage by a broader development of the fibrous band constituting the linea alba. This separation is always more marked at the intersection of the fibrous *lineæ transversæ*. In a hernial condition, the muscles are usually so far separated as to allow a suture to be placed at the borders of the aperture, without including the muscular tissue or opening their sheaths. In this kind of rupture, the size of the tumour bears even less proportion to that of its internal opening than is the case with the other varieties. I have frequently found, in very large herniæ, the internal aperture of no considerable size. This has more usually been the case where the sac contained omentum only. When bowel is present, the opening is usually larger. When the protrusion occurs through the true umbilical aperture, as in infants, the form of the opening is almost always circular. When it occurs in the linea alba, though close by the side of the umbilicus, as is generally the case in the adult, the opening may be oval or slit-like. This distinction becomes important both in the diagnosis and treatment of the disease in adults.

XXII.

DIAGNOSIS.

In the infant, umbilical rupture usually presents itself as a soft, rounded tumour at the site of the umbilicus, bearing on or near its centre the mark of the umbilical cicatrix. (Fig. 68.) It rarely exceeds in size that of a large chestnut. On pressure, it readily returns into the abdominal cavity, with a slight gurgling sound and flatulent feel, indicating presence of a portion of bowel. When its contents are returned, the smooth, circular edge of the tendinous aperture can usually be readily felt, and the skin and sac of the tumour inverted into the opening by the point of the finger, which can also, in cases of some size, be passed under the edge to some distance all round.

At this time of life there are very few or no conditions with which this rupture can be confounded. Fatty and other benign tumours are very uncommon in this situation, while malignant growths, springing from the root of the cord, are distinguished by well-known characters. The fungoid growth which occasionally arises from an unhealthy ulceration of the umbilical cicatrix, is usually easy to distinguish from hernia, because of the absence of its usual characteristics.

In some rare cases, shortly after birth, a distended or varicose condition of the umbilical vein remaining unobliterated up to the point of ligature of the cord, may in some degree simulate this hernia, and may predispose to it by keeping open the umbilical aperture. Nævoid tumours of this part are rare, and easily recognisable by their irreducibility, vascular, elastic dilating properties, and the absence of an opening in the tendon beneath them. When implicating the surface of the skin, the colour will distinguish them sufficiently. When subcutaneous and venous in character, a dusky or bluish colour given by the contained blood is usually seen through the translucent skin of the infant. In some cases I have observed such an appearance in a hernia also, caused apparently by the dark colour of the intestinal contents.

Fig. 68.

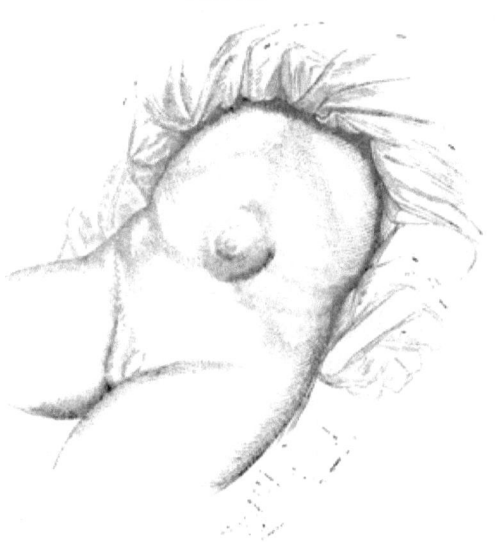

So rarely do other tumours which might simulate this disease occur in this part, and so easy is its detection and recognition when present, that no further remarks are necessary in this place in addition to the general signs of rupture detailed in the preceding pages. In the child it seems to give rise to but few symptoms. In an infant at present under my care, it is coexistent with a congenital inguinal rupture, neither of which seems of itself to distress the little patient, though they add considerably to its sufferings when labouring under intestinal irritation from other causes common to its age. In the adult, however, the symptoms are often very distressing, interfering much with the functions of the stomach and colon, with which its contents are closely connected, thus inducing an early accession of nausea, vomiting, and constipation, when any distension or obstruction occurs in the tumour.

At this time of life a fatty, fungoid, or recurrent fibrous tumour connected with the umbilical cicatrix, are the only diseases which may be confounded with this form of hernia. The manner of growth, steadily progressive, instead of alternating in size with the degree of general abdominal distension or disturbance, as is usually the case in hernial tumours; the absence of the usual effects of pressure in diminishing the size of the tumour to an appreciable extent, as we find in all cases of rupture not very extensively retained by adhesions; the evidence obtained from percussion and manipulation, which give a more or less resonant sound in all cases of enterocele, and a certain mobility, sliding, or yielding to the fingers, characterizing the contents of a hernial sac, whether omental or otherwise, all which are incompatible with the firm and fixed nature of the constituents of the above solid growths, especially in this part of the body; and the general appearance of the tumour itself, as well as that of the patient, together with the abdominal symptoms referable to the stomach, such as sickness, &c., or to the large intestines, such as constipation or other irregular action, will usually suffice for a clear diagnosis.

XXIII.

TREATMENT.

An umbilical protrusion in the infant can usually be cured by the application of a proper compress for two or three months. The strong tendency of the aperture to close by development, needs only that the viscera should be kept out of the sac to secure the gradual closing in of the edges of the opening. In many cases, however, this desirable result does not ensue, either from the application of a pad so convex or conical as to dilate the opening, into which it invaginates the integuments to as great an extent as the rupture itself; or from the great size of the aperture; or from a feeble power of development; or disease, sickness, and constant violent crying in the sufferer. If the hernia be left alone, or the means adopted fail to produce its cure, the longer its continuance, and the older the patient, the less likelihood is there of such a desirable result occurring by means of pressure only; since the propensity of the aperture to close grows weaker as the subject grows older. The incessant care and length of time required in the treatment by bandaging, and the difficulties of cleanliness under the circumstances, render these means more apt to fail in the children of the poor. It is said by Bichat that, after four years of age, the cure, either by pressure or operation, becomes difficult, and after nine years, impossible.

The operation referred to by this eminent surgeon is one performed by Dessault, as well as by himself, in numerous cases of umbilical hernia in children, and was simply a revival of the ancient practice described by Celsus. It consisted in pinching up the integuments and sac, and carefully squeezing out between the finger and thumb the contents of the latter; and then tying round the base of the tumour upon the surface many folds of ligature thread, tightening them from time to time until the included parts dropped off, leaving an open sore, which was afterwards treated by the application of lunar caustic and dry charpie.

The primary and most important objection occurring to this operation is based upon its insufficiency in securing the edges of the tendinous aperture, which were left untouched below the surface of the parts secured by the ligature. This embraced only the sac and integuments, as in all the earlier operations for the radical cure of inguinal rupture, described in the former part of this work. Most of the cases, consequently, demanded afterwards the application of a bandage for as long a period as is usually necessary for the cure by compression only; and one of the most important of the reasons for preference of the operation urged by Dessault, was thus rendered abortive. Such as were left without such treatment had, usually, a return of the protrusion afterwards. Another objection, urged by Cooper and other English and foreign writers, is based upon the supposed dangers of peritonitis; either primary, as the immediate result of the proceeding, or secondary, from an opening remaining after the slough dropped off, and communicating with the peritoneal cavity. This objection seems to have been founded rather upon the theoretically probable contingencies prevalent at that time upon the subject of operations upon the peritoneum, than upon the result of cases; but it may, at the present day, be held equally valid in the event of a prolonged opening into the cavity after the operation. A third objection, made on the score of the danger of including the bowel in the ligature, could only be referred, like others of the same kind, to the skill of the operator. A fourth, based on the consequences likely to result to the liver from its contiguity and continuity of surface through the umbilical vein and broad ligament, might with equal propriety be urged against the umbilical ligature applied at birth.

The testimony of the eminent Italian surgeon Scarpa is, however, strongly against the efficiency of Dessault's operation, as *never* producing a truly radical cure, as often producing dangerous symptoms, and as leaving an ulcer difficult to heal. The want of efficiency is the most fatal of these objections, and evidently proceeds from a failure in reaching the real seat of the disease—viz., the tendinous aperture. The dangers of the proceeding evidently accrue from the liability of the processes of sloughing and ulceration to precede, instead of following, that of adhesion of the apposed serous surfaces.

Of a still more severe and dangerous character seems to be the practice recently followed by some, of opening directly by incision into the sac cavity, paring the edges of the tendinous aperture, and then bringing them together by wire sutures. Such an operation has a character of severity, and involves apprehensions of danger from a direct exposure of the peritoneal cavity so near the more vital sympathetic nervous centres and abdominal organs, which the necessities of the case in adults, and still less in children, do not seem to require.

In infantile cases proving intractable to the treatment by pressure, fairly applied, which are few, but do occur; and in favourable cases in the adult, all of which have been hitherto pronounced altogether incurable, the adoption of the subcutaneous method of applying a wire ligature upon the principle laid down in these pages is likely to produce all the beneficial results obtainable by any operative procedure, without being exposed to the more real and reasonable objections fatal to the operations just mentioned. By this means the tendinous aperture can be reached without exposure of, and with very little interference with, the peritoneum; and its margins denuded and rendered adherent by the slow action of the wire ligature. The small size, great obliquity, and valvular nature of the punctures transmitting the wire, combined with the traction and compression of the loop formed by it under the skin, effectually prevent the admission of any injurious external influences to the cavity of the peritoneum, as will be seen on a perusal of the next section. The parts are thus placed, during the progress of the succeeding changes of ulceration and adhesion, under those favourable conditions which render subcutaneous operations generally much less liable to the risks which follow upon open wounds into important cavities or structures.

XXIV.

OPERATION FOR THE RADICAL CURE OF UMBILICAL HERNIA.

In three obstinate cases of umbilical rupture in children I have operated for a radical cure.

In the first case, I applied a small pair of rectangular pins vertically, and in opposite directions, one on each side of the opening, transfixing the edge of the tendinous aperture, and causing them to emerge close to each other through the integuments above and below the opening. When locked into each other, and twisted, the edges of the aperture were drawn close together in a line with the linea alba, protruding the integuments and sac of the rupture in a vertical fold above them. The pins were surrounded, and the skin protected, by strips of lint and plaster in the usual way, and were retained four days, producing no unpleasant symptoms by their presence. At the end of that time they were withdrawn, and the edges of the aperture were found firmly adherent. The child was afterwards dressed for a short time with a circular bandage and pad, and then discharged. While under observation, no return of the rupture had occurred.

Considerable difficulty having been experienced during this operation, in forcing the point of the pins through the very tough structures surrounding the hernial opening, I was led subsequently to substitute for them a tolerably long suture needle of the kind commonly used in operations, but stout, curved well near the point, and flat and broad at the eye; carrying a wire of the smaller calibre, but still stout enough not to cut through the tissues too rapidly. (Fig. 70, *d*.) To invaginate the coverings of the hernial sac, and to protect the bowel during the passage of the needle, a small spoon-shaped director (*a*), with a flat curved handle (figured edgewise in the woodcut), and a bowl-shaped scoop (*b*), large enough to fill up the hernial opening, is used. In small cases, an ordinary pocket-case scoop is sufficient for the purpose; or the smaller end of the director (*c*) may be used. The wire should be prepared in two pieces, eight inches long, bent into a small even hook at each end. It should be of copper wire, silvered, thick enough to hold without cutting its way, and flexible enough to follow the needle easily.

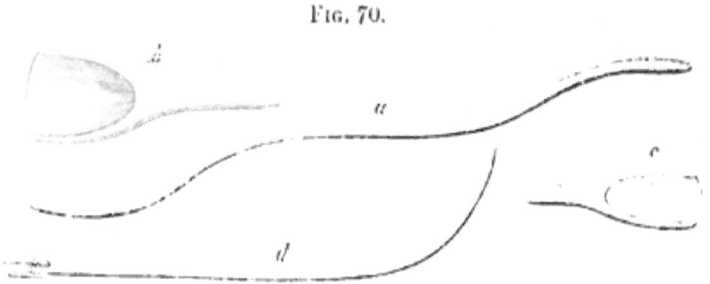

Fig. 70.

The patient being laid on the back, with the knees drawn up and the shoulders raised, the protrusion is carefully and completely returned. The convex surface of the bowl of the director is then pressed into the hernial opening, so as to carry the skin over the sac quite behind the edge of the tendinous aperture, on one side of the median line. The rounded end of the bowl must be pressed steadily and firmly against the deep surface of the tendon, pushing the skin as far as possible behind it. The needle, carrying one of the wires, is then placed in the hollow of the director, and its point pushed through the tendon from its deep surface, considerably above the centre of the opening. When the point is seen to raise the skin, the latter is drawn well upwards by the finger of an assistant, so that the skin may be pierced as far as possible *below* the aperture in the tendon. (Fig. 71.) The wire being drawn through and disengaged from the needle, the second

FIG. 71.

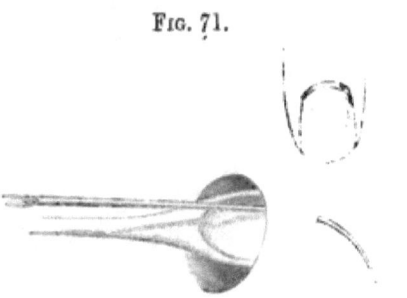

wire is hooked on, and passed in the same way through the lower part of the tendinous border on the same side; the skin being this time drawn downwards by the assistant, so as to enable the operator to pass the point of the needle through the same aperture as before made, or as close to it as possible.

FIG. 72.

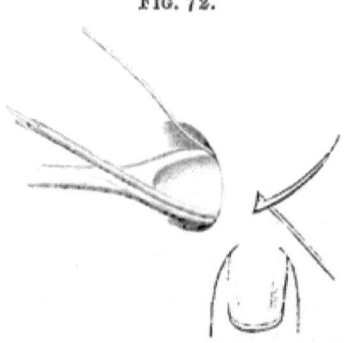

(Fig. 72.) The second wire is then drawn through, and the needle again disengaged.

The operator next directs his attention to the edge of the tendinous aperture on the opposite side of the median line; and, by repeating the same processes, he draws the opposite ends of the two wires consecutively through this border, at equal distances from each other, and through the same aperture in the skin.

In performing this manœuvre the director must be planted firmly within the hernial opening, between and behind the two wires which emerge at the centre of the sac coverings, (in the manner shown in Fig. 73,) so as to push the skin well over to the other side of the median line behind the edge of the tendon. The needle is then to be passed into the same apertures in the middle of the sac coverings at which the wires respectively emerge. (See also Fig. 74.) This can be more

Fig. 73.

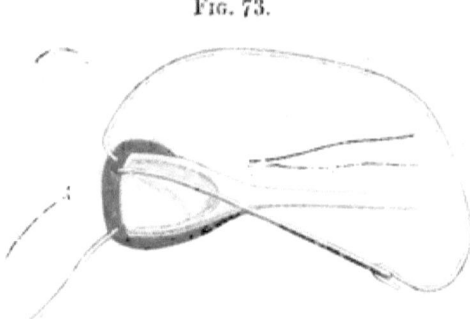

conveniently done before the bent ends of the wires are hooked on to the eye of the needle at each passage of the instrument. It is easy to accomplish it in this way unless the opening be very large, in which case it will be necessary to slide the point of the needle a little way close under the skin before piercing the tendinous structures, following the instrument closely behind by steady pressure of the end of the director.

The two wires are thus drawn across the hernial opening and through its tendinous borders, at equal distances apart, passing out at each side through the same puncture in the skin, and depressing under their pressure the sac coverings which lie in the area of the opening, at which point the wires, when drawn tight, disappear into the median punctures made for the temporary purpose of their application. (See Fig. 75.) If difficulty be experienced in bringing the needle through the same aperture in the skin at the sides each time, it may be overcome by making another temporary opening, drawing the wire through it, and then carrying it under the skin by passing the needle from one aperture to the other. If the hernial opening be very large, a third wire may be applied in the same manner across the centre of the opening between the other two.

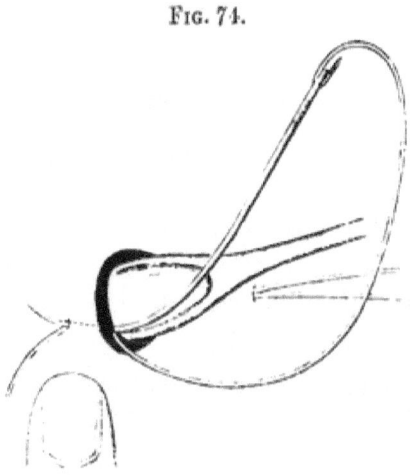

FIG. 74.

The ends of the wire on each side are then to be twisted up together into the puncture, until the hernial opening is felt to be closely shut up, care being taken at this stage that no protrusion of viscera occurs. This accident is, however, effectually guarded against, first by the pressure of the director in the area of the hernial ring, and then by the two wires passing transversely across the aperture under the skin. When the wires are firmly twisted up, the skin covering the sac is raised in a puckered fold in the centre, which is placed over the line formed by the approximated edges of the tendinous aperture hereafter to become united by adhesion under the pressure of the wire. (Fig. 75.) It will be seen, by referring to the figure of the parts in section (Fig. 76, *a*) that a portion of the tendon itself, close to the edge, is raised or everted under the pressure of the wire loop, so as to present a

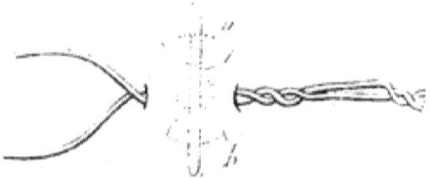

Fig. 75.

broader surface of adhesion than the mere border would afford. A very small portion only of the circumference of the tendinous aperture is left above and below the wires, if properly placed, as seen in Figure 75 at *a* and *b*.

The twisted ends of the wire being then cut off at a suitable distance, are hooked together over a roll of lint (Fig. 76, *b*)

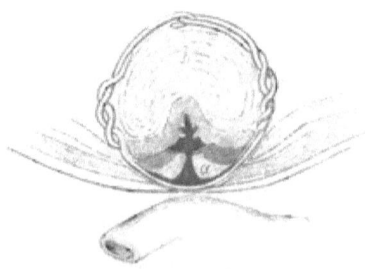

Fig. 76.

placed between them, so as to make pressure upon the puckered fold of skin and sac; the whole being retained by a strap or two of adhesive plaster and a circular bandage.

The patient should be laid on the back with the knees drawn up and the shoulders raised, and kept as free from crying and other exertion as possible.

In favourable cases of umbilical hernia in the adult this operation may be tried with but little danger, considerable probability of success, and the certainty of not rendering the state of things worse by interference. In many of these cases the danger of symptoms occurring is lessened by the possibility of performing the operation with very little interference with the peritoneal membrane, as by a dexterous application of the needle the edge of the tendon may be secured *obliquely*, without puncturing the peritoneum which covers it behind, because of the thickness of the former at this time of life.

The presence of much permanent distension of the abdominal cavity, by fat or other causes, I should consider altogether to preclude any attempt at a radical cure by operation, since a continuation or increase of the distension would probably bring about a return of the protrusion.

In a lax condition of the abdominal wall, and other favourable conditions, as of health, &c., being present, the adhesion of a portion of omentum to the parietes of the sac might not constitute a serious objection to the operation, since the presence of an omental plug would contribute to the closure of the aperture, and strangulation of this structure by the ligature might be avoided in its application. In the disposition of the sutures across the area of the hernial opening it will be observed that no implication of the omentum will be necessary under such circumstances, and no important strangulation of the protruded part will be possible, since the tension of the wires across it under the skin tends to force and keep back any protrusion, and the closer and more numerous the sutures, the less is the chance of such strangulation occurring. By this means, the opening in the tendon may be diminished very considerably, if not altogether closed, and a truss may thus be rendered much more effective.

XXV.

TREATMENT OF UMBILICAL HERNIA BY PRESSURE.

In the treatment of this variety of rupture by the pressure of a truss, the objections to a convex or conical truss pad exist to a still greater degree, especially in infants, where the opening has a normal tendency to close up under development.

The method commonly followed in the latter class of cases is to take a segment of a globe, or a graduated compress, composed of a substance more or less resisting, to press it into the aperture, and to retain it by plaster or a bandage round the body pretty tightly applied.

Such an apparatus must necessarily invaginate the skin and hernial coverings through the abnormal opening; and if it retain its position at all, dilate the aperture. If not very prominent, the convexity is rarely applied and kept upon the opening sufficiently accurately to accomplish its intention, and it thus becomes simply superfluous, if not obstructive. That a closure of the aperture often follows such an application only proves that a circular support and a certain degree of pressure only is necessary, in most cases, to prevent protrusion until the progress of development has accomplished a complete occlusion.

The application of the principles laid down in the foregoing pages to this subject has led me to employ an umbilical compress, in which the pressure is brought to bear upon the tendinous margins of the hernial opening in such a way as will tend to close them towards the centre or median line; thus aiding, instead of obstructing, the efforts of nature to contract the opening.

The best material for the compress is vulcanized india-rubber. A solid cylinder of this substance, about half an inch in diameter, is formed into an oval ring by the junction of its ends cut off square, leaving an aperture corresponding in size to the diameter of the hernial opening, and of a shape tending to an oval. Across one surface of the opening is then stretched a thin piece of the same substance, which is secured to the outer edges of the ring by an adhesive solution of the same material in chloroform.

Thus is formed an oval cup-shaped pad (*a*), with a valvular bottom fitting over the margins of the hernial opening.

Fig. 77.

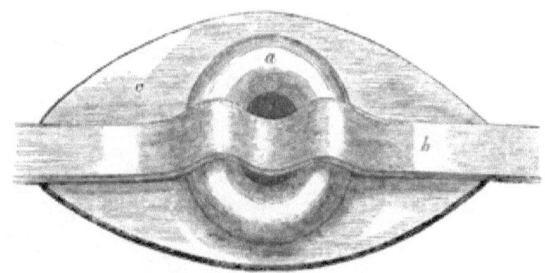

(Fig. 77.) A circular strap of the same material (*b*), large enough to enclose the body of the patient on a level with the navel, and strong enough to apply sufficient elastic pressure to retain the pad, is then fixed to the valvular bottom of the cup, on the hollow side, by attaching it to the surface of the valve by the adhesive solution, leaving the strap free and unattached as it passes over the rounded edges of the cup. Around the outer margin of the pad may be fixed a cushion of some soft material (*c*), with bevelled or sloping surfaces, to keep the pad more steady on the surface of the abdomen under the pressure of the strap.

When applied, the tip of the finger is passed into the cup-shaped depression of the pad sufficiently far to cause the valve to project at the surface. This projection is then fixed by the pressure of the finger into the hernial aperture after the protrusion is returned, and the elastic belt strap passed over the legs round the body of the patient. The surface of the skin over and around the navel should be first brushed lightly over with collodion, mixed with a little castor-oil, to protect it from chafing, and to make the surfaces of the valve and ring adhere more closely. When the pad is fixed with the long diameter of the oval placed vertically along the median line, the point of the finger is withdrawn, and the elasticity of the belt strap coming into play, the valve is retracted into the hollow cavity of the pad, drawing with it the skin covering the hernial sac which adheres to it through the collodion medium, and is thus drawn out of the tendinous opening through which the rupture protrudes.

The pressure of the elastic belt is thus brought to bear chiefly upon the thick margins of the cup, pressing them laterally upon the borders of the hernial aperture, and tending to close them towards the median line by pressure backwards and towards the centre of the opening. This lateral pressure brings the edges of the opening together from the side, producing an oval form corresponding to the shape of the opening in the pad, and excluding the bowel and viscera from the sac cavity, while the skin and hernial sac coverings are drawn out of the area of the aperture by the suction of the valve. In Fig. 78, (representing the apparatus and the parietes of the abdomen in horizontal section,) the dotted lines, $a\,c$, $b\,c$, represent the direction of the pressure traversing the edges of the tendinous opening (o). By this means all invagination of the skin into the umbilical aperture is avoided, and the pressure made entirely in the most effective manner—viz., upon the tendinous margins of the ring towards the median line. Protrusion of the viscera is effectually prevented by the closing in of the sides of the oval ring acting through the attachments of the skin on the margins of the tendinous opening. The slight concavity which is produced by the suction action of the valve is entirely filled by the skin and sac covering, and does not admit of the protrusion of bowel.

For hospital purposes, a good substitute for the india-rubber ring is gutta percha. The valve may, under these circumstances, be made of a broad strap of common adhesive plaster placed across the skin surface, and bent over the edges of the ring into the bottom of the cup surface thus formed, to which it may be made to adhere, and afterwards be carried by the

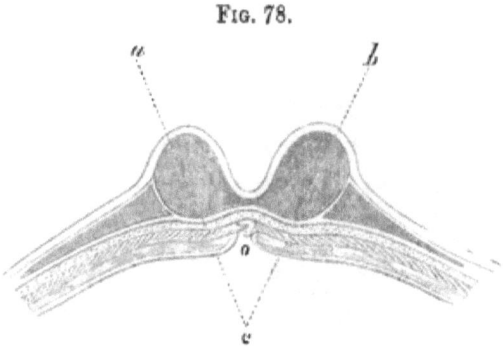

Fig. 78.

long ends to encircle the body, so as to fulfil the part of the circular belt, the whole being secured by a bandage. A little collodion brushed on the surface of the skin, and the reverse of the adhesive plaster which is applied to it, will preserve the former from friction, and cause the latter to adhere. Pieces of wash-leather, of diminishing sizes, laid over each other, with a hole cut in the centre to receive the gutta percha ring, will serve the purpose of a cushion around the pad, and keep it steadier under the motions of the body. If considered advisable, a piece of tape or ribbon may be used instead of the adhesive plaster, which, in children with a delicate skin, is apt to irritate, and always causes the adhesion of dirt, though it possesses the advantage of more close adaptation to the surface.

In the cases upon which these apparatus have been tried the results have been very satisfactory, producing a more rapid closure of the umbilical aperture than I have ever obtained by the old method recommended by Sir Astley Cooper and others; in which the compress is both difficult to retain " in situ," and when fairly held in position, resists instead of promoting the remedial processes of nature.

The same apparatus may be used with advantage in the hernia of adults. If a portion of omentum be irreducible, the opening of the pad may be made of sufficient size to include it. The pressure is thus brought to bear upon the margins of the opening around the protruded and irreducible portion; and may, by being persevered with, set up adhesive action around it, so as to prevent further increase of the tumour.

APPENDIX OF CASES.

RADICAL CURE OF INGUINAL HERNIA.

Case 1.—John Craig, æt. 25, a printer, applied at the Lincoln's Inn Dispensary with a right, direct, inguinal hernia, of eighteen months' duration. Has had during that period frequent obstructions of the bowels, and was once on the point of being operated on for strangulation, but the rupture was finally reduced by the taxis. Has had seven trusses, none of which retained the rupture for any length of time; the last giving rise to abscesses in the groin. The hernia filled the scrotum, sometimes attaining the size of two fists. On reduction, the deep opening was found large enough to admit the tips of three fingers, with lax, loose margins. On the slightest cough or exertion, the bowel dropped into the scrotum. The patient being willing to submit to an operation, it was performed in April, 1858. The method in which the operation was done was that first described, by ligature thread and compress. The instruments used were, however, somewhat different. Instead of the finger, a curved tube mounted on a handle was used to invaginate the fascia, a needle of like curve being passed through it carrying the thread. The use of this tube was afterwards discarded, greater confidence being felt in the tactile perception of the finger. The compress used consisted of a quadrangular piece of box-wood, with a hole in the centre to transmit the ligature to the cross piece of wire on which it was fastened. On the third day after the operation, the incision in the scrotum was healed. On the 5th, the compress and ligatures were removed; the upper puncture suppurating freely. In a fortnight this also was completely healed. No pain in the belly or testis was complained of, and the patient was out of bed in rather less than three weeks; when he was exhibited in the theatre of King's College Hospital. The inguinal canal was then felt to be obliterated by a band of adhesion uniting the pillars of the superficial ring, with much consolidation along the canal and at the deep opening. No impulse whatever was felt on coughing; the groin being more firm and resisting than that on the opposite side. This patient wore a truss for some months after the operation, and as the newly-formed tissues contracted, a swelling appeared in the scrotum which had all the appearance of a varicocele, with some effusion in the fundus of the sac. The truss was thereupon discontinued for a while, and the enlargement then slowly disappeared. During the first few months after the operation, the patient underwent a severe test of the cure in the shape of a severe bronchitis, which has occasionally affected him ever since. After seven months the use of the truss, which had been resumed, was entirely discarded. His occupation has necessitated the frequent lifting and removal of heavy "forms" of type, which often weigh as much as a hundredweight. This patient was among those exhibited to the Fellows of the Medico-Chirurgical Society, on the occasion referred to in the text, and since that time he has been frequently exhibited in the theatre of King's College Hospital. On the last occasion (a few months ago), the groin operated on could not have been distinguished from the other, except by the presence of the small scars. No impulse whatever, or any bulging was observable.

Case 2.—William Mayhew, æt. 54, a sailor, was on board the *Dreadnought* hospital ship, under my friend Mr. Tudor. He had a direct scrotal hernia on the left side, of six years' standing. It had been strangulated three or four times, and narrowly escaped operation. He has worn out as many as thirty-three trusses, which have, latterly, failed in retaining the bowel. Wurtzer's operation for the radical cure had been, a few weeks before, performed on the rupture, but on his again assuming the erect posture, the bowel has redescended to as great an extent as before. With the kind permission of Messrs. Busk and Tudor, I operated on this patient (though from its size and character it was far from being a hopeful case) April 6th, 1859. The forefinger only was used to invaginate the fascia and protect the needle. Two ligature threads were employed, passing through separate openings in the skin. The deep opening and superficial ring being very large and lax; the canal very short; and Wurtzer's operation having previously interfered with the parts entirely unsuccessfully, I was not very sanguine as to the result of the case. On the 9th, pus having begun to show at the openings, and appearing to be retained somewhat by the lower ligature, this was withdrawn. There was no sickness or tympanitis, the pain and tenderness being confined to the groin. On the 11th, the remaining ligature and compress were withdrawn; discharge free; bowels open naturally. On the 15th, he was not so well, having caught cold during the severe weather which then prevailed; tongue white; appetite bad; feels low; wounds looking well, and discharging freely good pus. Had been ordered brandy. On the 20th, much better; feels all right again; much less discharge; wounds closing up. On the 13th of May, the patient showed himself at my house, with the groin entirely healed, and much hard consolidation apparent in the canal. A hard, cord-like mass of transplanted fascia passed into and obscured the superficial ring, and was lost in the general consolidation of the passages. When the patient coughs, there is less impulse than on the opposite side, which is rather weak and bulgy. This patient has shown himself at various times at King's College Hospital. When I saw him, four months after the operation, the groin was still hard and firm, and no impulse was apparent. He was still wearing a light truss. He then went to sea, and I heard several times from his son (also a hernial patient of mine) that he continued well, and able to do duty. Last summer, however, he called upon me with the rupture returned to some extent, though not so large as before the operation. He states that it reappeared gradually some months before, after performing very heavy labour in whipping coals with an inefficient truss, on board the collier ship on which he was doing able seaman's duty.

The next two cases were operated on by me at Fort Pitt, Chatham, May 11th, 1859:—

Case 3.—A sergeant of foot, æt. 45, tall, and very muscular, with a right oblique and scrotal rupture, of seventeen years' standing, with a long and narrow inguinal passage. During the operation, notwithstanding a fair amount of chloroform administered, the recti muscles acted so powerfully, that some difficulty was experienced in passing satisfactorily the needle through the conjoined tendon. Two ligatures were applied, one close up to the internal ring, and the other close above the pubic crest; all the ends being brought through the same cutaneous aperture, in the skin of the groin. The large, square, boxwood pad was employed. On the sixth day, the ligatures and compress were removed, sufficient action being then set up, with very little constitutional disturbance. The wounds healed satisfactorily. When last seen, a slight fulness in the groin was the only evidence of the previous rupture.

Case 4.—The son of a sergeant, æt. 8 years, with a left, oblique, inguinal rupture, probably congenital, but only first observed when the child began to walk. One thread suture only was applied, the sides of the canal coming together very readily. Ligature and compress removed on the sixth day. Two days afterwards, a nettle-rash appeared on the skin, succeeded by a smart diarrhœa, which continued two or three days. On the twentieth day after the operation, the punctures were completely healed, a complete cure of the rupture being effected, with no impulse or fulness in the groin. This condition continued unimpaired to the end of June, when the lad embarked with his parents for India.

The particulars of the two last cases were afforded me by Surgeon-major Matthew, under whose care the patients were placed. This gentleman has himself operated many times, after my method, with success. In one case, however, he was unfortunate enough to have erysipelas in the ward at the time the case was there, and the patient died from inflammation spreading along the sheath of the rectus muscle, which seems to have been more interfered with in the operation than I have been in the habit of doing.

Case 5.—Ignatz Gutler, æt. 31, a compositor, was operated on in the theatre of King's College Hospital, May 14th, 1859, for a large oblique, scrotal hernia, on the left side, of two years' standing. States that, of late, it has been unusually painful. The operation was done in the way described in the text, with thread and compress. The day after, he complained of much pain in the testis, along the cord, and in the groin, which was relieved by opium. There was no sickness, tympanitis, nor abdominal disturbance or tenderness. On the 17th, a swelling was apparent in the epididymis and cord, which was evidently mainly owing to enlargement of the spermatic veins, with some œdema. There was thirst, irritative fever, white tongue, and increased pain. Under these circumstances, I thought it judicious to remove the pressure, though earlier than usual. The lower or scrotal incision had then nearly healed, but still discharged a little; the discharge of pus from the upper puncture being copious. Next day, the patient was much relieved, and consolidation to a considerable extent was apparent in the canal. 19*th*.—Bowels opened by castor-oil; less swelling; no impulse on coughing. From this time he daily improved, the swelling gradually diminishing. On the 28th, he was exhibited in the theatre of the hospital, with the lower wound entirely, and the upper nearly closed; the varicocele having nearly disappeared. The superficial ring could not be felt by the finger, and there was no cough impulse whatever. On the 8th of June, he left the hospital, wearing a light truss. About a month after this, he sent for me, and I found that the swollen condition about the cord and sac had returned, larger than before. He stated that the swelling had come on without pain or exertion, and while still wearing the truss; the bowels were confined. A brisk aperient was prescribed, and he was directed to discontinue the truss, as tending to increase the swelling. A week after this, the patient was again exhibited at King's College Hospital. Though made to cough freely, no impulse or impression could be distinguished at the superficial ring, by careful examination with the finger. A hard, round, elastic tumour could be felt amid the enlarged veins, which, from its weight and obscure fluctuation, seemed to be an effusion of serum into the fundus of the sac, cut off from the peritoneum by adhesions in the canal. The difference between the feel of the tumour and that of a rupture was evident. After that, I saw the patient twice, and on repeated examination fancied that there was possibly a small channel of communication with the peritoneum still existing, as the fluid could be squeezed slowly out of the distended sac fundus, with a sensation exactly similar to that given by an india-rubber bottle. At that time, there was clearly no return of the rupture, though not having lately seen the patient, I am disposed to set it down as a doubtful case, in which the distension of the effused fluid might open up the canal afresh, so as to admit of a protrusion of the viscera.

Case 6.—Thomas Dyer, æt. 58, a labourer, operated on in King's College Hospital, June 11th, 1859, for a right, oblique, scrotal rupture, about the size of an egg. Has had the rupture five years; it has latterly been getting larger, and more difficult to retain. It has been twice strangulated, the last time a month before, at which time the bowels were not relieved for the space of a week, notwithstanding large doses of purgatives. In operating on this case, I remarked that, though the superficial ring was barely large enough to admit the finger, the canal itself and the deep ring were much larger than the external appearance indicated, so that it would have been impossible to distend the interior of the canal by the use of Wurtzer's plug. The day after the operation, a severe and troublesome bronchitis, to which he had been subject, came on, the cough being so violent that I had fears of the ligatures giving way under it; this, however, did not happen. The compress and threads were removed three days after the operation. Much hard swelling was observable in the canal; the discharge from the upper puncture free; bowels opened by castor-oil six days after. In this case I removed the ligatures so soon, rather experimentally, to ascertain the amount of action absolutely required to produce the desired results. Notwithstanding the bronchitis, the patient was out of bed on the twelfth day after the operation; and on the seventeenth the wounds were healed. He wore a weak truss for a short time, and then growing impatient, he discontinued it, and went to harvesting, doing a great deal of heavy lifting entirely without truss support. On showing himself to me in January, 1860, I was gratified to find that the adhesions were still resistant. The patient expressed himself very confident of a cure after so severe and long-continued a test without any return. In February of the same year, he was shown at the meeting of the Medico-Chirurgical Society. There was then a slightly suspicious impulse when the patient coughed hard. This impulse became more marked each time I saw him, at different intervals, not being able to persuade him to resume his truss. The last time I saw him, in the summer of 1862, there were decided evidences of the rupture having returned to some extent, though it was much smaller than when operated on. The case may, therefore, be looked upon as a failure, owing to three causes, viz., the short time during which the ligatures were kept in, the occurrence of the bronchitis, and the want of subsequent truss protection during heavy labour so short a time after operation.

Case 7.—William Mayhew, æt. 20 (the son of Mayhew, of *Case 2*), operated on in King's College Hospital, June 17th, 1859, for a left oblique, scrotal rupture, congenital, and for which no truss had ever been worn. The tumour attains sometimes the size of a large hen's egg, drops easily into the scrotum, and is reduced without difficulty. On the night after the operation, had a morphia draught, but did not sleep. Next morning had some sickness, which was relieved by an effervescing draught. On the 20th, complained of much pain in the groin; no tympanitis, or tenderness, on pressure over the belly; sickness better. To have pil. saponis co., gr. v. every three hours, if the pain continues. 21*st*.—Pain much less; sutures and compress removed, sufficient consolidation being apparent in the canal. Free suppuration in the track of the ligatures; abdomen quite soft and compressible; tenderness limited to the groin. A small slough has appeared at the upper margin of the scrotal puncture, apparently produced by the pressure of the end of the compress. Ordered wine ʒiv. daily. 24*th*.—Free suppuration from both openings, slough separating. 30*th*.—Both wounds fast closing in; slough gone; scrotum well tucked up to the superficial ring; consolidation very apparent in the upper part of the canal; no impulse felt on coughing. *July* 13*th*.—Patient discharged cured; the upper puncture being entirely healed, the lower nearly so. Has been kept longer in the hospital on account of a diarrhœa, with which he was attacked, and which also affected, to a greater extent, two other cases operated on about this time, being prevalent in the hospital in the autumn of that year. (*Cases* 10 and 11.) This patient has never worn a truss, either before or since the operation.

He was exhibited at the Medico-Chirurgical Society, Feb. 28th, 1860; having been employed since the operation in the laborious work of piling shells in Woolwich Dockyard. At that time the groin was remarkably firm and resisting, no impulse whatever being felt in the canal. A slight bulging over the deep ring was the only evidence remaining of the original deficiency in the groin. I have since repeatedly seen, and have heard of this case up to a recent period, through his father and other hernia patients, whom he has sent for a radical cure. He still continued sound and without a return of the tumour. I look upon it as a fair test of what can be done without the assistance of truss support, after a radical cure has been attempted by operation.

Case 8.—Jacob Lorge, æt. 27, a German baker, operated on in King's College Hospital for a right, oblique, scrotal hernia, of fifteen years' duration. In this case, the scrotum was remarkably pendulous, reaching eight inches down the thigh, from the combined effects of hernia and varicocele. The sac was remarkably elongated, and rested on the testis, which hung much lower down on the side affected than on the left side. The veins of the cord were very large and tortuous, and there was much loose tissue and fascia about the scrotum. Operation performed June 29th, 1859. The scrotal opening was made lower down than usual, to get the finger well behind the sac, carrying it well up into the deep ring. The patient was rather more restless and feverish than usual, for the first few days after the operation. On the evening of July 2nd, the scrotum had become so much swollen, tense, red, and painful, that the compress and ligatures were removed. From both punctures a reddish serum oozed copiously. Along the canal, and at the superficial ring, a large hard mass could be felt, which appeared at first like a return of the protrusion, a supposition rendered more probable from the patient having caught a smart cold, and coughed very vigorously for the last two days. There was, however, no distress, sickness, or vomiting, and the tenderness on pressure was confined to the groin. There was no tympanitis whatever in the belly. The mass felt solid, and could not be compressed or returned into the canal. Fomentations, poultices, and pil. saponis co., gr. v. every night. On the 4th, the bowels were well opened by castor-oil, and suppuration freely established in the punctures. The hard swelling was more apparent and pronounced, and was evidently solid effusion. On the 8th, the scrotal enlargement was much diminished; a very free suppuration, mixed with numerous white sloughs, issued from both punctures, especially the lower, proceeding evidently from the interior of the sac, by the walls of which they were apparently formed. On the 12th, sloughs in considerable quantity continued to escape, and the scrotum was reduced to one-fourth of its former size; a hard mass blocked up the inguinal canal entirely. On the 16th, the scrotum was reduced to less than its size before the operation. Numerous healthy granulations were apparent at both wounds. On the 17th, the patient got out of bed; and on the 23rd, he was discharged, with the wounds reduced to a very small size. He was afterwards exhibited in the theatre of the hospital, and the extraordinary effect of the operation upon the length of the scrotum and the varicocele excited universal remark. The scrotum on the side operated on was reduced in length one half, the right testicle being drawn up much higher than the left. The varicocele had entirely disappeared. A hard, solid mass blocked up the entire inguinal canal, and no impulse whatever was felt on coughing. When I last saw this patient, he had worn no truss whatever since the operation, and expressed his intention of visiting his native country, and showing himself to several eminent German surgeons, who had pronounced him incurable. He promised to let me know faithfully if the rupture returned; and as he has not done so, I conclude that the case remains still, as it bid very fair to be, an excellent cure.

Case 9.—John Batson, æt. 17, operated on, June 29th, 1859, for a left oblique rupture of three months' duration, descending into the upper part of the scrotum to the size of a small egg. This case went on with remarkably little distress, the patient hardly losing one night's sleep, with no opiate after the first night. The compress was removed on the fifth, and the ligatures on the ninth day after the operation. On the eighteenth day, the discharge (which had been pretty copious) had nearly ceased. The scrotum being well tucked up, and adherent to the superficial ring, the patient was allowed to move about the ward. His general health had been excellent, his appetite good, and his bowels regular throughout. After the sixth day, he had meat diet and wine. On the 19th of July he was ready to be discharged, but was kept to the 23rd, to be shown in the theatre of the hospital. The wounds were then almost healed, the groin hard, firm, and resistant, and no impulse was felt at the superficial ring, which was so blocked up by adhesions as scarcely to be distinguished. He wore no truss after the operation. On the 15th Oct., 1859, he was exhibited at the hospital, having gone through his work, as an engineer in Woolwich Dockyard, without truss. He was among the patients shown at the Medico-Chirurgical Society, in Feb., 1860, having worn *no truss* whatever since the operation. The groin was then very firm, the superficial ring closed up by adhesions, and no hernial impulse could be felt to distinguish one side from the other. A few months ago I again saw this patient. After some very heavy work with the hammer, and lifting hard, he felt a slight pain in the part, and applied to me for examination, having worn no truss up to that time. No appearance of rupture was present, but, on coughing hard, a weakness was very evident over the deep ring, and a looseness of motion in the parts, which led me to recommend the adoption of a truss during the hours of labour. The patient himself felt no inconvenience beyond that above mentioned.

Case 10.—John Laramy, æt. 17, a pale, sickly-looking lad, was operated on in King's College Hospital for an oblique scrotal rupture on the right side, of sixteen months' duration. The case was remarkable as presenting the largest internal opening of any I had operated on, admitting with ease the tips of four fingers, with a very lax condition of the surrounding tissues. Had never worn a truss. Operation done July 13th, 1859. From the width of the openings, the case was like one of direct rupture, and the strain upon the ligatures was great. Being apprehensive of their retaining power, I kept on the spica bandage and linen compress longer than usual. This, I believe, had an injurious effect in retaining the discharges. On the 16th, the compress was removed, the threads being left in two days longer, to lead off the discharges, which had begun to appear copiously at both punctures. He then complained of colicky pains in the belly, for which pil. sapon. co., gr. v. was ordered thrice a day, with hot fomentations and a poultice to the wound. 17*th*.—Complains still of pain, flatulence, and distension of the abdomen, for which he was ordered a draught of tinct. opii, tinct. rhei. co., sp. æth. chloric. and aq. menthæ, thrice daily. 18*th*.—A copious discharge of thin, gruelly pus, chiefly from the upper puncture; pain and flatulence easier, but still present; tongue white and furred. Strong beef tea ad lib.; milk diet. 19*th*.—Discharge more copious; abdomen more distended, and harder; tympanitic, but not tender on pressure, except over the groin; no sickness; pulse 120. To have two drachms of castor-oil, and a large poultice over the groins and stomach. 20*th*.—Bowels opened four times with the oil; belly very tympanitic; no sickness, nausea, or vomiting; tongue covered with thick white fur; pulse 120; discharge still very copious, thin, and gruelly. To omit the mixture; to take pil. sapon. co., gr. v. every four hours, and to have 4 oz. of wine daily. Turpentine stupes to upper part of belly; continue poultice to the groins. 21*st*.—Bowels open four times in the night; symptoms continue unchanged. 22*nd*.—The distension of the belly continues, but no vomiting or sickness is present, nor is the tenderness on pressure marked; a little is felt along the iliac and lumbar regions of the side operated on; tongue very foul; pulse 100. A large enema of the confection of rue, with gruel, was now administered by means of a long rectum tube, with the effect of bringing away a large quantity of flatus. Repeated the next two days with the same effects, giving much relief to the patient, and diminishing the distension to a remarkable degree. Pulse 98; discharge thicker and more creamy, still very copious, but less in quantity; the punctures remain widely open. During the four following days the bowels were much relaxed, but his appetite recovered, and his looks were more cheerful. Sleeps well without the pills, which are now discontinued. 29*th*.—The diarrhœa continuing, he was ordered decoct. hæmatoxyli ʒj, tinct. kramerioe ʒss, after each loose evacuation. 30*th*.—Diarrhœa checked; tongue cleaner; pulse 98, soft and compressible; discharge again more copious. Pressure forwards along the iliac crest induces a free evacuation of matter from both the upper and lower punctures. The pus seems to have burrowed between the muscles in this direction. *Aug.* 1*st*.—A drainage tube introduced into each opening, and passed upwards towards the iliac crest. This giving rise to pain and uneasiness, was next day removed; the escape of matter is favoured by turning the patient on to his left side. A large sponge kept to the wound to absorb the discharge. His appetite is now very good: to have a slice of meat, potatoes, and half a pint of stout. Nitric acid, chloric ether, and cinchona mixture, thrice daily. From this time the discharge gradually lessened, and the patient slowly gained strength. Much consolidation was apparent in the canal. On Sept. 14th, he was able to walk about, the lower opening being entirely healed, the upper one still discharging a little healthy pus. No impulse whatever was felt on coughing; the groin was hard, firm, and resistant, the hardness reaching upwards and backwards along the iliac crest and lumbar region; wears no truss. Sent to Margate for a change of air, after having been in hospital just two months. *Nov.* 26*th*.—The patient was shown in the theatre of the hospital, having worn no truss since the operation, four months ago. Has quite regained his health and strength; rupture quite cured; no impulse whatever. The superficial ring cannot be distinguished. *Feb.* 28*th*, 1860.—Shown to the Fellows of the Medico-Chirurgical Society, having worn no truss until within a day or two of this date, when he was recommended to apply one on account of a slight fulness which had become apparent in the groin. There was no impulse whatever on coughing; the bulging limited to the centre of Poupart's ligament, where also the abdominal cough impulse was limited at the upper cicatrix. Feels the groin strong and firm. When last seen, the patient continued in the same satisfactory state, and was about to tramp for work as a shoemaker.

Case 11.—John Albert, æt. 42, a healthy, muscular man, operated on in King's College Hospital, July 23rd, 1859, for an oblique, scrotal rupture, on the left side, brought on by lifting nine months before. Had worn a truss, which did not prevent the bowel from slipping into the scrotum, which it distended to the size of an egg. The compress and ligatures were removed three days after the operation, when the lower opening was found nearly healed, the upper one discharging moderately. The same night he had sickness and diarrhœa, then prevalent in the hospital. Next day the abdomen was flatulent and distended, but not painful on pressure; respiration abdominal; tongue white and foul; pulse 100. Had been sick, and vomited, and had the bowels purged three times in the night. Discharge free and thin. Two days after, the flatulence still continuing, but without further purging, an enema of rue and gruel was administered, giving great relief, and causing much discharge of flatus. The diarrhœa afterwards returned, and continued for three or four days, notwithstanding the administration of logwood decoction and tinct. krameriæ; these medicines, however, had at length the effect of checking the diarrhœa, when the tympanitic distension gradually disappeared. During this time the skin and subcutaneous structures around the groin puncture had assumed a brawny hardness and a dull reddish hue, with a thin, purulent discharge. As he recovered from the diarrhœa this gradually disappeared under the use of poultices, leaving a small collection of matter above the scrotal puncture, which was evacuated by opening the latter up with a probe. On Aug. 10, he was discharged, cured, having been about the ward some days with the punctures almost healed; there was no impulse whatever. The patient left the hospital wearing no truss. He was enjoined, and promised, to apply again, or write, if the rupture returned. Not having done so, it may be concluded that he has no trouble with it.

Three of the cases just given were in the hospital at one time during an autumnal season, which was very sickly. They were affected with very similar symptoms of flatulent distension and diarrhœa, which was prevalent in the house, and so may fairly be attributed to this accidental cause. Partly, if not chiefly, to the lowering effect of this epidemic diarrhœa, may be attributed the copious suppuration and retarded recovery of *Case* 10. In this case a burrowing of matter gave rise to peritonitis, affecting chiefly the parietal layer, and secondarily, perhaps, the lower part of the small intestines; and therefore not giving rise to the stomachic distress usually observed in this disease. The unhealthy constitution of the patient, and the great strains upon the ligatures, from the large size of the hernial openings, seemed to contribute to this untoward result, leading to a prolongation of the period of recovery nearly three times the usual extent.

Case 12.—George Jolly, æt. 15, a sailor boy from the Marine Society's ship, operated on at King's College Hospital, Oct. 15th, 1859, for a right, oblique rupture, distending the scrotum to the size of a small egg, of four weeks' duration, and leading to his discharge from the ship. One suture of ligature thread only was applied, and fastened over a wooden compress. On the third day the thread was tightened up by twisting round the pad. On the fifth day they were both removed. The pain was confined entirely to the groin; belly quite natural; discharge trifling in amount. On the 27th, the upper puncture entirely healed; and on the 29th, the lower also, at which date he was discharged, wearing no truss. A fortnight after he was exhibited in King's College Hospital, the groin being then in a satisfactory state, and no cough impulse being felt. A week after this, he was foolish enough to lift a hundredweight without the support of a truss, and thereupon felt a sudden pain and a sense of giving in the groin, attended by a return of the rupture. Not being then able, as he wished, to have the operation repeated at once, it was postponed from time to time until the rupture, in spite of the truss he had resumed, got larger than before, and more difficult to be retained. At length, after two years' time, he was again admitted for operation, which was done on the improved method by the twisted wire suture, Oct. 5th, 1861. On the 6th, it was found that he had contracted a gonorrhœa; for this he was ordered frequent syringing of the urethra with warm water, an effervescing draught every four hours, and pil. sapon. co., gr. v. at bedtime every night. 8*th*.—Complains of occasional pains in the belly, and scalding of the urine. No tenderness on pressure, sickness, or tympanitis; pulse 80; tongue white; bowels not open. Castor-oil ordered. 9*th*.—Bandage removed. No pain or swelling in testis or scrotum, wound looking very well; little discharge from puncture; free discharge from urethra. 10*th*.—Bowels opened with considerable griping; complains of the scalding and increase of gonorrhœal discharge; no tenderness on pressure except around punctures; belly quite soft and yielding; punctures rather red and obstructed by inspissated discharge; some healthy pus escaped on moving the wires; a round mass of induration at lower puncture; no swelling or tenderness of the testis on either side. To have a draught of liq. potassæ, tinct. camph. co. and sp. æth. nitric., with camphor mixture, thrice daily. *12th.*—Feels much better; less scalding; redness at punctures disappeared; discharge lessened. *15th.*—Doing exceedingly well; tongue clean; appetite good; bowels open. Induration in canal very evident. Gonorrhœa better. *19th.*—Going on very well; wires withdrawn, easily and with little pain (fourteen days after the operation); no bleeding followed. *22nd.*—Lower wound entirely healed; upper shows healthy granulations with serous discharge; gonorrhœa nearly well; dress with compress and spica bandage; to get up to-morrow. From this time to Nov. 2nd he was about the ward, with a small opening only present in the groin. He was at this date shown in the theatre. No hernial impulse whatever was present. Superficial ring obstructed by adhesion, and a hard mass felt in the upper part of the canal. He was discharged from the hospital without truss, and walked about without one until Nov. 18th, when he came to me with one on. The groin was then all that could be desired, firm, hard, and resistant, with the ring scarcely to be distinguished by the finger. In the autumn of 1862, he was shown in King's College Hospital, having remained well, and the groin giving no trouble. The external ring could now be felt more distinctly, but would not admit the tip of the finger; no hernial impulse whatever was to be felt; still wears the truss when at work; promises to come directly any change is evident to him.

Case 13.—J. F., æt. 35, operated on July 23rd, 1859, for a small inguinal hernia on the left side, which had previously been twice operated on, once by Wurtzer's method, and once by Gerdy's; the result was, a very patulous condition of the superficial ring and canal, admitting the forefinger with ease. A tumour with a cough impulse filled the canal, appearing at the ring; a thin, lax condition of the abdominal wall, and a sense of great weakness and pain in the groin on making any muscular exertion. The patient stated that his condition was quite as bad as before the above operations were done. In this case I operated with the threads and compress. They were removed on the third day, and the patient was out of bed on the ninth day; the consolidation in the canal along the track of the ligature being very satisfactory, and the closure of the superficial ring entire. A month afterwards, a little bulging only was apparent in the groin opposite the deep ring. To give this support a light truss was recommended. In the September of 1861, I was accosted by this patient to thank me for the excellent cure which had resulted. He had thrown off his truss entirely, and the groin was everything he could desire. This was more than two years after the operation. Some months ago, I again saw this case, at which time there existed a decided tendency to a return of the rupture, as shown by the impression produced by the cough upon the superficial ring. I therefore recommended the resumption of a proper truss. He stated that the alteration appeared after making unusual muscular efforts without truss.

Case 14.—J. S., æt. 38, a healthy, fine-looking man, who had come from California to be operated on for the cure of a very troublesome and painful rupture of the right side, of twelve months' duration. Stated that he was operated on in America soon after the tumour appeared, by a seton drawn through the sac. Describes the operation as very severe and painful, keeping him in bed a long time, and producing much suppuration. As soon as he began to move about, the rupture reappeared, although a truss was constantly worn after the operation. The cicatrices in the groin and scrotum were very extensive, and the external ring very patulous. The neck of the sac is wide, its direction being more vertical than usual. The rupture descends directly into the scrotum, attains the size of an orange, and has not been kept up by any truss that he has tried. As the anatomy of the canal has been much interfered with, I was not very sanguine as to the result of the operation, but, at the patient's earnest request, I operated on the 17th Nov., 1859, with ligature thread and compress. Some difficulty was experienced from the cicatrices, in invaginating the fascia to the requisite extent, and in detecting the conjoined tendon. The sac was funnel-shaped, with a large internal opening, the finger moving freely within the dilated canal. To avoid the old cicatrices, the scrotal puncture was made lower down than usual, and the puncture in the groin close to Poupart's ligament. This I afterwards found to have the effect of bringing the pressure of the compress too low, so that it did not bear satisfactorily over the axis of the hernial canal. The compress was removed on the 5th, and the thread on the 6th day, not a disagreeable symptom having supervened. On the 28th he walked about the room without truss. In a day or two, after the wounds were completely healed, and on the 6th Dec., he walked out of doors without truss. Soon, a certain mobility was apparent about the fascia attached to the superficial ring, which, on pressure being made at the upper cicatrix, was drawn a little upwards into the canal. Ten days afterwards, he went to Geneva, whence he wrote, a month after the operation, to say that the hernia had returned, expressing at the same time a great desire for another operation. Accordingly, he returned to London, and was again operated on January 21st. This time the punctures were made directly through the old cicatrices; the thread and compress were again used. On passing the finger into the canal, I could distinctly feel the fascia which had been transplanted in the first operation, adherent to the cord and anterior wall. The hernia had descended to its inner side, where the mobility of the fascia had first appeared after the first operation The position of the conjoined tendon was not at all evident, the width of the internal opening having apparently almost obliterated it. The compress was removed on the fifth day, the lower ligature kept in until the thirteenth, and the upper till the eighteenth. In the meantime, he had got up and walked about the room. Very little pain, and not an unpleasant symptom had occurred. Not much action was set up in the canal, and the discharge was very trifling. The speed with which reparation and absorption proceeded in this case was remarkable, as well as the little effect induced by the presence of the threads. On the 13th of Feb., 1860, both wounds were closed, a hard, firm mass of consolidation apparently filling up the canal. The scrotum was shortened, but the testis was moveable, and free from enlargement. He went again on the Continent, wearing a light truss. On returning, after a few weeks' absence, he called upon me previous to sailing for California. There had been great absorption in the canal, and the induration was nearly all gone. The superficial ring was much narrowed, and the point of the finger could not be inserted; there was, however, some impulse on coughing, an impression upon the walls of the canal rendering the case not entirely satisfactory. The truss, however, was rendered perfectly available for supporting the weak groin, which had not been the case before the first operation. I recommended him to wear the truss constantly for a year. Having then some intention of again visiting this country, he promised to call again on me and report progress, but has not yet done so.

Case 15.—Ann Martin, æt. 29, a healthy, unmarried woman, in service, applied with an oblique inguinal rupture on the right side, of four years' standing, which no truss that she had procured was effectual in retaining. The rupture was of the size of a fist, and seriously interfered with her means of livelihood. Operated on at her own home Feb. 17th, 1860, by ligature, thread, and compress. In this case the stitches were placed external to the sac, which was pushed up by the invaginating finger. From the absence of scrotal fascia much difficulty was experienced in getting far enough within the canal; the labial fascia not being divided resisted the invagination, and the hold of the threads upon the conjoined tendon was consequently doubtful. For the first few days she complained of much pain in the wound, and was hysterical.

The belly was soft and compressible, and no tenderness except in the groin. Compound soap pill, gr. v., every night. On the 23rd, the compress was removed, and a poultice applied; suppuration moderate; consolidation slow in forming. *March* 8.—Ligature threads removed. On the 12th, she walked to my house with the wounds entirely healed. There was then no impulse on coughing, and she felt much stronger in the groin than before the operation. She wore no truss. In about a month she came again to me, and I found that the rupture had reappeared to some extent, but not so large as before the operation. Recommended to wear a truss constantly, in the hope that continued pressure might still produce contraction of the canal, after the action that had already been set up. Since then I have not seen her.

Case 16.—Stephen Larkins, æt. 43, of Plumstead, a tall, stout, healthy looking excavator, with a double scrotal rupture, produced by a fall of earth, almost burying him alive, four years ago. Both are of the oblique variety, the left one much the larger, attaining an enormous size, equal to two fists, and admitting easily the tips of three fingers through the deep opening. The right side gives him less trouble, the truss being effective in keeping up the rupture, which is not the case with the left side. Wished the left side to be operated on, as he could manage the other with a truss. Operated on at King's College Hospital, May 16th, 1860, by threads and compress. Two ligatures were applied, one close to the deep opening, and the other near the pubis. Much strain was apparent when the threads were drawn close. On the 17th, he had a slight retching, and was very restless, complaining of much pain in the groin, which was allayed by repeated doses of sol. of morphiæ hydrochlor., $\mathfrak{m}$xx. every three hours. His face was flushed, and he perspired freely from the effects of the morphia and pain. Respiration abdominal throughout; no tympanitis or tenderness on pressure over the belly, except in the groin operated on; pulse 60; bandage removed. 18*th*.—Did not rest much last night as the pain returned towards evening, after having entirely abated. Had a little nausea; slept pretty well all the morning; no tympanitis; tongue white; pulse 60. Pil. sapon. co., gr. v. every four hours; fomentations. 19*th*.—Much better; slept during the night; no nausea or vomiting; much less pain in the groin; tongue a little furred, but moist; pulse 64. To leave off the pills. 23*rd*.—Compress removed today; suppuration free; bowels open for the first time since the operation; pulse 60. 28*th*.—Discharge diminishing; much induration evident in the canal and ligature tracks. On June 2nd, the wounds being much contracted, he got up and walked about the ward. The scrotal puncture healed entirely next day. The threads were kept in the upper wound until June 8th, and on the 11th the puncture closed. On the 16th, the patient was exhibited in the theatre of the hospital; a great deal of induration was evident in the canal; the scrotum was well tucked up, and firmly adherent by its cicatrix to the pillars of the superficial ring, which were firmly closed. No impulse whatever was perceptible, and the contrast with the opposite side was very remarkable. I have since had a note from the wife of this patient, in answer to my inquiry, stating that he continued to be in a satisfactory condition, and intended coming some day to see me.

Case 17.—George Vernan, æt. 21, of Chatham, operated on at King's College Hospital, June 2nd, 1860, for a right, oblique, scrotal rupture, of two and a half years' duration. When lifting, the rupture slips behind his truss into the scrotum, and gives him much pain. The operation was done with thread and compress. Next day complained of much pain in the groin; had not slept since the operation, in spite of the usual dose of opium; tongue furred; mouth parched; feels feverish; no tenderness on pressure except in groin; pulse 88. 4*th*.—Pain decreased very much; felt comfortable; slept well; pulse 95; tongue furred; skin rather hot. 5*th*.—Fever gone; very little pain; begins to eat; pulse 72. 6*th*.—Compress removed; free suppuration. 8*th*.—Bowels moved; pulse 68; sleeps without pills. 12*th*.—Could eat more; looks thin, pale, and anxious; pulse 76, not resisting; ordered full diet; nitric acid and bark mixture. 13*th*.—Free suppuration; testicle swollen, and a little tender on the side operated on; to be supported and fomented. 14*th*.—The pain in the testis prevented rest last night; pulse 72. This pain in the gland was easier the next day, was attended by very little enlargement, and did not excite further notice during his stay in the hospital. *June* 21*st*.—Threads withdrawn; much induration. 23*rd*.—Walked about the ward. 26*th*.—Discharged; external ring firmly closed; no impulse whatever; testicle still a little tender when pressed. *July* 3*rd*.—Exhibited in the theatre of the hospital; scrotum on the side operated on much contracted; testis drawn up towards the ring, and about one half the size of its fellow; canal and ring firmly closed by cicatrices; no impulse whatever on coughing. Since that time I have once seen this patient many months after the operation. The cough impulse was equal on both sides; the wasting of the testis not greater than before. I have since heard of him; he continued quite free from the rupture.

Case 18.—Bartholomew Connor, æt. 23, a sailor, operated on at King's College Hospital, June 16th, 1860, for an oblique, scrotal rupture in the right groin, brought about by an accident on board ship a year and a half before. Operation by thread and compress. On the

18th, had not had much pain since the operation; slightly feverish; pulse 80. 21st.—Bowels open naturally; slight tenderness. 22nd.—Suppuration established; did not sleep well last night; compress removed; lower opening healed; thick pus squeezed out of the upper puncture. 23rd.—Pain and feverishness gone; begins to have an appetite; a good deal of thick matter squeezed out of the upper puncture; poultice. 26th.—Less discharge. 28th.—Much induration apparent. 30th.—Ligatures removed; to get up and move about the ward. *July 7th.*—Discharged; canal and rings firm and contracted; went out without truss. Three months afterwards this patient, having been at work as a common sailor, without truss, since his discharge, returned to the hospital with a small tumour, which proved to be a return of the rupture, to about half the size it was before. The superficial ring admitted the point of the finger. Not having at the time a bed vacant in the hospital, the patient's wish for an immediate repetition of the operation could not be acceded to, and he was told to apply in a week's time or so. He did not do so; and I heard afterwards that he had been operated on again on board the *Dreadnought* by Mr. Tudor, after my method, successfully.

Case 19.—James Mead, æt. 25, a grocer, sent to me for operation by Mr. Coulson, for a right, oblique, scrotal rupture of a year and a half's standing, upon which Wurtzer's method had been unsuccessfully tried. The rupture had returned as large as ever one month after this operation, and had since then become more painful, the truss failing to retain it. Operated on at his own home, June 25th, 1860, by thread and compress. The anatomy of the canal was rather obscured by the previous operation. 26th.—Has had a good deal of pain; bandage removed; no tenderness anywhere but in the site of the operation; pulse 84. 27th.—Slept better; compress removed; suppuration commencing. 29th.—Much less pain; more discharge; belly quite soft and natural. *July 3rd.*—Bowels opened by castor-oil. *9th.*—Doing very well; ligatures removed; suppuration moderate. 11th.—Got up to-day. 13th.—Came down stairs. 15th.—Resumes his occupation in the shop. 17th.—Upper wound quite healed; lower nearly so. 21st.—Shown to-day in the theatre of King's College Hospital, not having worn a truss since the operation. Superficial ring quite closed; canal firm and resisting; impulse the same on both sides. To wear a light truss when at his employment. *May 30th,* 1861.—Came to-day to the Lincoln's Inn Dispensary; after wearing the truss four months has left it off entirely; the impulse, on coughing, equal on both sides, each presenting a slight bulge over the centre of Poupart's ligament; superficial ring does not admit the top of the finger. This patient has, since this time, been many times exhibited in the theatre of King's College Hospital, and the parts have remained without alteration. The last time he was seen was on Aug. 29th, 1861, when he continued in a most satisfactory condition, doing occasionally much lifting, and wearing no truss.

Case 20.—Richard Cook, æt. 39, of Woolwich, married; operated on in King's College Hospital, July 31st, 1860, for a left, oblique, scrotal rupture, of six months' duration, by the thread and compress operation. He had afterwards the usual amount of pain in the groin; belly quite free from tenderness and tympanitis; no sickness; going on quite in the usual way; no swelling of the testis. 24*th*.—Compress removed; wounds look healthy; discharge healthy and moderate; bowels not yet open; complains of cough, thirst, and pain in the chest; pulse 90, full. To have a draught of liq. ammon. acet., chlorodyne, and sp. æther. chloric., with aq. camph., every four hours, and sinapisms to chest. 26*th*.—Much better; cough relieved; appetite pretty good; pulse 80, compressible. 28*th*.—Cough nearly gone; tongue slightly furred; punctures looking well; suppuration pretty free; consolidation can be distinguished. 29*th*.—Bowels acted freely. 31*st*.—Ligatures removed; discharge healthy; pulse 70. *Aug.* 1*st*.—Expresses himself as feeling extremely well; appetite and pulse good. 4*th*.—Bowels acted freely from a dose of castor-oil. 9*th*.—Has gone on very well since last report; wounds gradually closing, and discharge decreasing in quantity; lower puncture entirely healed, upper diminished to a sinus, discharging healthy matter; much consolidation in the canal, and less impulse than in the opposite groin; stood up to-day at the bedside without pain or inconvenience, except from weakness. Up to this time there had not been, in this case, the occurrence of a single bad symptom; and at this date I left town for a short time, with directions for his discharge as soon as the sinus was healed. For two days after this he was up and about the ward. He was then seized suddenly with a smart rigor, succeeded by feverish symptoms, foul tongue, and quick pulse. Dr. Johnson and Dr. Evans saw the patient, and he was put on ammonia, brandy, quinine, and strong beef tea. In a day or two, the symptoms became those which characterize blood-poisoning—viz., very quick pulse, recurring rigors, profuse perspirations, dusky countenance, and diarrhœa; the small sinus still remaining ceased to discharge, and assumed a dark livid hue. Further notes of the case were unfortunately not kept, it being the vacation. After a week's absence I returned to town, saw the patient, and carefully examined the groin, perineum, and abdomen. *Aug.* 18*th.*—There was no tympanitis; the belly was soft; no swelling, redness, nor tenderness whatever on pressure about the wound; the canal was hard, firm, and resisting; the groin puncture was reduced to a small fistulous track, admitting a probe to the depth of three-quarters of an inch downwards and inwards; it was dry and glistening. There was felt a slight, dull pain, on pressure above the pubis into the pelvis. No pain or difficulty was felt in passing the urine, which was rather high-coloured, but in proper quantity; slight pain felt, on pressure in the perineum, by the side of the urethral bulb. On seeing him next day, I found the symptoms decidedly aggravated; the rigors were more frequent and severe; the perspirations profuse, with a sweetish faint smell. He had low muttering delirium, and a dusky, sunken face; tongue covered with a dry brown fur; pulse 130, and very compressible. There was now some tympanitis, and rather more tenderness on pressure. He had taken 16 oz. of brandy during the twenty-four hours. He sank and died, Aug. 21st, 1861.

Necropsis.—On opening the abdomen, there was found no effusion of fluid into the peritoneal cavity, nor was the smooth glistening surface of the intestines altered, but the ilium was more vascular than normal, indicating commencing peritonitis. There was no effusion of fluid or pus in the sheath of the rectus, or in the course of the spermatic vessels, or vas deferens of the side operated on. All these parts were perfectly normal. A small fistulous track led from the puncture into the inguinal canal, and was arrested by the posterior wall. The parts about the conjoined tendon were much thickened, and consolidated into a hard, fibrous mass, adherent to the border of the rectus muscle. On section of this indurated tissue, there exuded from small points on the cut surface (apparently sections of veins) a small quantity of purulent fluid. The peritoneum covering the bladder on the *opposite* side of the body was dark and congested, but still smooth on the surface. On raising this, and dissecting into the vesical fascia, the tissues were found hard, and resistant to the knife, and an ounce or so of thick pus flowed from the section. On closer examination of the cut tissues, the pus seemed to have been contained, not in a single cavity, but in canals of tortuous course, which were evidently the dilated and hypertrophied veins of the vesical plexus between the bladder and pubis, extending down towards the triangular ligament and perineum. This deposit evidently accounted for the deep-seated pain, on pressure in the perineum, observed two days before death. It was remarkable as extending more on the opposite side of the pelvis than towards the side operated on. Behind the external pillar of the ring operated on, a small channel still remained unclosed by the ligature, in the neck of the hernial sac, extending along in front and to the outer side of the spermatic cord. Its upper and inner boundaries were composed of the thickened and puckered sac, thrown into longitudinal folds, and matted together by adhesive inflammation. The diameter of the remaining canal would admit only of a goosequill. Its upper, posterior, and anterior boundaries were hard and unyielding. Its lower side was the spermatic cord, and its serous covering of sac little altered. At the deep opening of the hernial canal the omentum was found adherent to its upper boundary for about one-half its extent, and covering the greater part of the puckered opening. The gut in immediate proximity was not more vascular than that which lay in the pelvis, indeed, hardly so much so. The liver was large and light-coloured, probably fatty; and the kidneys were large and flabby. The mucous surfaces of the bladder and ureter were normal. Further examination was prevented by the wishes of the friends.

This unfortunate case bid fair both in its progress up to the occurrence of the pyæmic symptoms, and from the condition of the parts operated on, as found after death, to have been a very satisfactory cure. The symptoms and post-mortem appearances clearly indicate that the fatal result was owing to those, as yet inscrutable causes, which are quite independent of the position, nature, or severity of an operation, and result in a purulent infection of the blood. Although, in the present case, the quantity of suppurative action was by no means so great as in many which preceded and did perfectly well, yet the occurrence of a fatal case led me to turn my attention to some method of diminishing, if possible, the amount of suppuration in the progress of the case, and to the adoption of wire as the agent for drawing together the sides of the canal. I had before remarked that the amount of adhesive action was not generally proportionate to that of the discharge; that the former appeared to depend more upon the vigorous condition of the patient's health at the time of operating, and the length of time during which the ligatures were kept in; and the latter, upon the violence of the action set up and the degree of tension exercised upon the sutures in bringing the sides of the canal together.

Case 21.—William Freeman, æt. 21, operated on in King's College Hospital, Sept. 29th, 1860, for a left, oblique rupture of twelve months' duration, brought on by lifting, in the course of his duty as a sailor. This case was remarkable for the presence of two internal openings into the inguinal canal—one superficial and external, in the site of the deep ring; and the other smaller, through the conjoined tendon, in the triangle of Hesselbach. The hernial tumour protruded through the former opening, and a cough impulse could be detected in the latter also. The operation was done with threads and compress; the lower end of the ligature was carried across the face of the lower deep opening, so as to include its borders in the grasp of the ligature; the upper end of the ligature crossed the front of the upper opening or internal ring. A broad compress was used. During the two following days he had pain in the groin and scrotum of the side operated on, but no other unpleasant symptom, and, on the whole, expressed himself comfortable. *Oct. 2nd.*—The compress was removed (three days after the operation). The swelling and tenderness in the groin was very slight; no abdominal symptoms whatever; a small quantity of healthy pus flowing from the upper puncture; lower nearly closed. Compress reapplied, and ligatures tightened over it to produce more action. *4th.*—Had a restless night; more tenderness and swelling in the groin; discharge scanty; bowels acted freely; not much pain; enjoys his food; compress removed altogether. *6th.*—Comfortable; bowels open; discharge natural. *10th.*—Doing well. *16th.*—Discharge much diminished; consolidation very evident in the canal; has been walking about the ward. *18th.* Ligatures removed; wounds closing up rapidly. *20th.*—Wounds quite healed; much induration in the canal; superficial ring quite closed; less cough impulse than on the opposite side. This patient was discharged, wearing no truss, but was recommended to get one on returning to his employment, and to let me know if any return of the rupture occurred. He has not hitherto done so.

Case 22.—Henry Holland, æt. 22, an engineer, operated on in King's College Hospital, Oct. 6th, 1860, for a left, direct, scrotal rupture, with a very patulous and lax canal and rings, of two years' standing. The operation performed in this case was the first in which wire was used, still retaining the use of the compress. On account of the great size of the hernial openings, the two ends of the wire, after being passed through the conjoined tendon and Poupart's ligament close to the deep opening, were crossed over each other in front of the orifice, and again passed through the pillars of the ring, as described at page 125, and shown in Fig. 37. By this means the patulous superficial opening was closed effectually, close above the pubic crest. Next day he had passed a good night; pulse 82; tongue a little white. *11th.*—Some œdema of scrotum apparent; a little pain on pressure of the testis; no suppuration; compress removed. Has had a troublesome cough, for which a mixture was ordered; poultice to the groin and scrotum.

13*th*.—Bowels open by castor-oil; punctures discharging a little; no abdominal swelling or tenderness; tongue white; pulse 90; cough still continues. 16*th*.—Œdema of scrotum still continues; no pain in the testis; cough and feverishness better; wires removed without difficulty; poultice to the groin. To the 2nd Nov. he went on rather slowly but gradually progressing; there was but little discharge; the lower wound was then nearly healed, but a thickening of the scrotum in front of the cord still remained. He then complained of pain in the scrotum, which became more swollen and tender at the part where the fundus of the sac lay. Suppuration in the sac being anticipated, fomentations, poultices, and a dose of castor-oil were ordered. Next day the lower wound began to discharge more freely, and during the next three days the swelling and tenderness abated; and the testis could be felt a little enlarged, but not tender. On the 6th he was all right again. 10*th*.—Wounds nearly healed; up and about the ward. 15*th*.—Discharged; groin firm, hard, and resisting; no impulse on coughing below the upper cicatrix; testis and cord a little enlarged, but not painful or tender; to wear his truss when walking or at work, but not at other times. *Feb.* 12*th*, 1861.—The patient came to me with the parts but little altered; there was, perhaps, a little more bulging above the upper cicatrix, but the testis and cord were diminished in size to nearly their normal dimensions; still wears a truss when at work. *April* 30*th*.—Again saw this patient; a slight protrusion can be felt high above the pubis, to the inner side of the upper puncture, which has a slight impulse on coughing; the superficial ring below this point can be felt quite closed, and not admitting the point of the finger; no impulse on coughing can be felt in this position; testis and cord of their normal size. The small protrusion seems to be either through the puncture made by the needle in the tendon, or through the upper part of the hernial opening. The truss was adjusted upon the weak part, and he was recommended to wear it constantly, and to apply again if the tumour gets larger, and does not disappear. I have not since seen the patient.

Case 23.—David Mumford, æt. 21, pale and sallow, a sawyer in Woolwich Dockyard; operated on in King's College Hospital, Oct. 13th, 1860, for an oblique, scrotal rupture, which had been previously operated on for the radical cure by another surgeon, without success, the rupture returning when the patient began to move about. The failure was apparently owing to the conjoined tendon not having been properly secured. In this case I operated with wire in the manner described at page 123, and shown in Fig. 36, being anxious, as the rupture was of considerable size, and had returned after one operation, to secure fully the lower as well as the upper part of the canal. The opposite groin was weak and bulgy; the hernial openings opposite to each other; and the whole make of the individual such as described in the text under the second class of hernia cases. The boxwood compress was used also in this case. Next day, had slept well after the usual pill; no abdominal symptoms; no pain except from the pressure of the compress; pulse 80. *15th.*—Very little pain; a little suppuration apparent. *16th.*—A slight swelling of the scrotum; doing very well. *18th.*—Complains of cough; a little feverish; no pain in belly or groin; bowels open by castor-oil. *19th.*—Cough more troublesome, giving pain in the groin; to have a cough mixture. *20th.*—Cough better; suppuration increased; compress removed. From this time he went on very well, having but little suppuration from the punctures, and much induration in the canal and track of the ligatures. On the 30th, the wires were removed without difficulty by dividing the upper loop, and untwisting and cutting off the lower ends close to the wound, and then drawing the two portions upwards separately. On Nov. 2nd, he was up and about the ward; the punctures closing up with remarkably little suppuration; testis and cord normal. On the 9th, he was discharged, with the punctures healed; the superficial ring and canal quite closed and firm, and no cough impulse below the upper cicatrix. He went out without truss, but was recommended to wear one when at work. On Jan. 20th, 1862, this patient again applied at the hospital with the hernia returned, but not to its former size. States that, after wearing the truss three or four months, he discarded its use altogether, as it was in the way of his work, which consisted in pushing heavy trucks along with his belly, causing the truss to hurt him. After going on with this work without truss for four months longer, the rupture began to reappear, and speedily obtained its present size, he having worn no truss since. To resume his truss. Stated his resolution to have the operation repeated at some future time, but has not since applied.

Case 24.—Captain B——, æt. 26, a military man, operated on for a right, oblique, scrotal hernia, of six years' standing, which gave him great trouble from its constantly slipping down under the truss. Being about to go out to India, he was anxious to have it permanently cured. Operated on March 30th, 1861, in the manner shown in Figs. 33, 34, and 35, at pages 120 to 122. At night the pulse was 80; no fever; had a sedative draught, and slept very well. Next day expressed himself as very comfortable; pulse 80; tongue clean; little or no pain. *April 2nd.*—Suffers very little; sleeps well; bandage removed.

An extra twist given to the wire over the compress to set up more action; lower puncture nearly healed; very little action; no tenderness anywhere. Bandage replaced firmly, so as to make compression. *3rd.*—Feels very little inconvenience; tongue clean; wishes for more food; glass of sherry daily. *4th.*—Restless night, but no pain; a little matter oozing from the puncture; some incrustation over the lower one; opened it with the probe, letting out a little pus; more action in the canal; slight œdema of the scrotum. *5th.*—Bowels open by castor-oil. *6th.*—More pain and discharge in the punctures; thickening more evident, with slight cutaneous redness; compress removed; poultice. *9th.*—Discharge now tolerably free. *12th.*—Discharge lessening; lower opening a mere fistulous track; water dressing and pressure upon the wires "in situ" by compress and bandage. *15th.*—One half the wire removed, the other left to keep up a slow indurative action. *17th.*—Got up to-day; stood erect without pain; lower puncture closed; discharge small, entirely serous; no impulse felt on coughing below the wire; cord and testis quite free from swelling or tenderness; the hernial sac seems to have been entirely drawn up into the canal. On the 20th, thinking that not quite enough action had been induced, I reapplied the boxwood compress over the single wire which remained, and applied firm pressure by a spica bandage. *24th.*—The increased pressure having set up a considerable amount of hard thickening, the remaining wire was removed, after some difficulty, from a kink which had formed upon it, being retained by the granulations. Serum only has been discharged from the punctures, no pus having appeared during the last week, although moving about with the bandage on. Compress and bandage lightly reapplied. *30th.*—Wounds almost entirely healed; no cough impulse felt along the canal; superficial ring completely blocked; can bear the pressure of a light truss; to go out of doors. *Aug. 27th.*—Called to-day to take leave previously to his departure for India; continues in a most satisfactory condition, wearing the truss only when walking or riding; the external ring can be a little more plainly felt; no cough impulse whatever; a hard cord can be felt in the canal, which he states to become a little more prominent when he sneezes or coughs sometimes. This I could not detect. Is in capital health, and feels much stronger than before the operation. Promises to write if he has any reappearance of the rupture.

Case 25.—Thos. Rogers, æt. 23, a healthy, fresh-looking young man, with a double inguinal rupture. That on the left side is large and scrotal, and gives him much inconvenience; that on the right side is a small bubonocele, projecting halfway down the canal. The groin and genitals have a general fœtal outline, as described in the third class of cases in the text; the lower part of the internal oblique is deficient in development. Operated in King's College Hospital on the larger rupture, March 16th, 1861. The operation done in this case was that described at page 126, by means of wire held fast by a steel clamp. *17th.*—Had a good night; very little pain; tongue clean; pulse normal. *18th.*—Doing very well; bandage removed; upper puncture closed by adhesion; very little discharge from the lower; testis a little swollen and tender; some effusion into the fundus of the sac in the scrotum; no abdominal pain or tenderness; tongue a little furred; pulse 80. *19th.*—Doing well; sleeps without sedative; little discharge; wires tightened one turn; water dressing. *23rd.*—To-day the steel clamp was removed, the wire being left in; more discharge; a good deal of consolidation, which extends more than usual into the scrotum; the fundus of the sac is hard and swollen. *24th.*—Bowels freely opened by aperient pills; no pain in belly or testicle. *26th.*—Complained of pain in the groin, at the site of the upper cicatrix, where a little induration and some œdema can be felt. Fomentation and poultice to the groin. *27th.*—Pain and swelling of groin increased; indistinct fluctuation; the swelling in the scrotum is larger, and is evidently due to effusion into the sac fundus, being distinct from, and to a certain extent moveable upon, the testicle. Tongue furred; feverish; pulse 108. An incision made over the cicatrix in the groin, and some thick pus evacuated. Wire withdrawn from the lower puncture by traction on one end, some force being necessary to accomplish its removal. *28th.*—Much better in every respect; swelling in groin nearly subsided; no tenderness or tympanitis over the belly; tongue white; pulse 100; more free discharge from the punctures. *29th.*—Swelling in the groin disappeared; free discharge of thick pus; pulse 90. *30th.*—A small black slough apparent within the scrotal opening, lying upon the cord, and formed evidently by the fundus of the hernial sac; discharge free; tongue clean; pulse 80. *April 1st.*—Much improved in every respect; inguinal canal much indurated; dark-coloured, thin discharge from scrotal opening. *12th.*—The slough from the scrotal opening removed, being about the size of half the little finger; upper wound nearly closed. *18th.*—Rather more swelling and discharge from the upper part of the scrotum, proceeding from the presence of more sloughing of the sac; the testicle is painful when pressed, not otherwise, and gives its peculiar sensation to the patient; groin hard and firm, and the neck of the sac evidently completely obliterated. *24th.*—Had more pain in the scrotum, at the lower part of which is a soft, red, fluctuating swelling in front of the testis. This was opened, and a quantity of grumous pus and shreds of slough were squeezed out; the testis could afterwards be distinctly felt, rather diminished in size, but of normal shape and feel; could eat more; to have two eggs daily; has been taking strong beef tea; milk diet; and brandy, ʒiv. daily. *27th.*—To-day, a slough two inches in length, and of the diameter of the little finger, was drawn through the scrotal opening, formed evidently by the remainder of the sac. *30th.*—The swelling in the scrotum has entirely subsided; discharge thick and yellow, from the fresh opening in the scrotum. Both punctures granulating well; testicle can be distinctly felt, a little wasted, but giving its normal sensation to the patient. Injection of zinci sulph. Feels quite well; middle diet and half-a-pint of porter. *May 7th.*—Testis can now be more clearly felt; is smaller than the other, but still tolerably plump; to get up and walk about the ward. From this time the punctures gradually closed up, and the patient was discharged on the 22nd July, wearing a light double truss. The superficial ring on the side operated on perfectly closed and firm; no cough impulse whatever; a slight bulging apparent at the site of the deep ring; the testis is about half the size of its fellow. The patient expressed himself entirely satisfied with the result of the operation. I have seen him again a short time since, and find the groin quite resisting; no return of the hernial protrusion, but with the slight bulging before alluded to; the testis is shrunken to a fourth of its natural size; still wears the double truss. I attribute the trouble in the scrotum in this case, and the wasting of the testis, chiefly to the pressure of the steel clamp, and partly to the healing of the counter opening in the groin. The abscess in the groin was clearly due to the closure of this aperture; the fever and excitement set up by it in the parts, and the increased pressure upon the cord, aggravated the action in the sac already set up by the presence of the clamp. The result was sloughing of the sac, and so much interference with the supply of blood to the testis, as to result in atrophy. Since this case, I have always taken care to keep the groin puncture open, and have substituted twisting of the ends of the wire for the use of a clamp.

Case 26.—Thos. Taylor, æt. 24, a seaman, H.M.S., operated on in King's College Hospital, April 6th, 1861, for a left, oblique, scrotal hernia. This patient had been twice before operated on by another surgeon, by simply invaginating the skin of the scrotum into the canal, and holding it there by sutures passed through the pillars of the ring, by a method closely resembling Gerdy's. The second opera-

U

tion, however, had been modified by making a scrotal puncture and tucking up the fascia only, an attempt having been made to include the conjoined tendon also. Directly after the first, and six weeks after the last operation, the rupture returned into the scrotum, where it now formed a tumour as large as the testicle. He had been much troubled by a severe bronchitis, and when he first applied to me for a third operation, was expectorating freely, and had much pain in the chest. Consequently I strongly recommended him to wait until his cough had been cured. Being very anxious to get back to his ship, he, a short time after this, represented himself as free from cough, and ready to undergo the operation. This was afterwards found not to have been the case, as in the early part of the treatment he suffered much from the effects of the cough upon the groin. The operation done in this case was the wire operation without compress, described in the text (page 109). A good deal of hard, thickened cicatrix was felt on the posterior surface of the outer pillar of the ring, evidently the result of the former operations. The rupture had returned behind these adhesions. The wire in the operation was placed above and beyond these, and passed fairly through the conjoined tendon, upon which a distinct pull could be felt when the wire was twisted up. The sac in the scrotum was very thick, tough, and resisting to the needle, when passed across it in front of the cord. On the night of the operation, the cough began to be troublesome, which he explained by saying that he had got a fresh cold. Next day, the expectoration was copious and bronchitic. He complains much of pain in the groin when he coughs, but not at other times; tongue covered with thick, white fur; pulse 100. Bandage and dressing left undisturbed, and a draught of liq. ammon. acet., sp. æther. chloric., with aq. camph., ordered every four hours, with pil. sapon. co., gr. v. every night. 8*th*.—Cough very violent; expectoration very copious; complains of much pain in the testicle; scrotum red, and the gland exquisitely tender and rather enlarged; pulse 110; tongue much furred; face hot and flushed; dressings removed, upper puncture not at all inflamed; no swelling in the canal; no pain, tenderness, or tympanitis in the belly; respiration abdominal, both flanks rising and falling equally. Scrotum to be well supported and fomented; poultice to the groin; turpentine stupes to the chest, and the pills three times daily. At ten P.M. much easier; no pain in the testicle; sleeps much; pulse 108. 9*th*.—Very much better in every respect; swelling, redness, and tenderness of scrotum and testis much diminished; cough and expectoration less; no abdominal pain or tenderness; pulse 100. 10*th*.—Continuing to improve; moderate thick discharge from the punctures; tongue cleaner. 11*th*.—Continues better; bowels not open since the operation; pulse 96; to have an ounce dose of castor-oil. 13*th*. —The oil not having operated, a common enema was given, producing two copious stools; cough better; tongue slightly furred; pulse 94; wounds look very well; discharge thick and moderate; testis less tender. Light pressure over the wire loop was made by straps of adhesive plaster, so as to make pressure against the posterior wall of the canal. 15*th*.—Doing well; cough still troublesome; testis slightly enlarged and still a little tender. Water dressing and spica bandage over wire loop. 24*th*.—Cough quite gone; bowels confined; eats well; wishes for ale. To have a pint daily; wire untwisted but not withdrawn; wounds granulating; discharge thick and moderate; testicle and scrotum have resumed their normal appearance and size. 30*th*.— Wires withdrawn (under chloroform, at the patient's request); no difficulty experienced; the ends were moved freely in the wound, and cut off close to the skin; the loop above was next drawn upwards and extracted entire. *May* 1*st*.—Punctures closing in; much consolidation in the canal and upper part of scrotum. To get up and move about the ward with a spica bandage and lint compress firmly applied. 4*th*.—To-day the patient was shown in the theatre of the hospital. Punctures entirely healed; groin firm, hard, and resisting; superficial ring perfectly closed; no cough impulse whatever; condition most satisfactory. Discharged; recommended to have a light truss when at work; promised to let me know if there be the least change or return of the rupture; I have not heard from him to the present time.

Case 27.—Alfred Spencely, æt. 8 years, operated on at King's College Hospital, Feb. 22nd, 1861, for a left, oblique, scrotal rupture, which had been observed one month; the superficial ring was large, and cough impulse extended. In this operation, the rectangular pins were used for the first time, as described in the text, and given in Figs. 40 and 41. Next day, the little fellow had some difficulty in passing his water, as it seemed, from fear of hurting his groin. Complained of no pain anywhere; belly soft and pliant; pressure does not cause him to shrink, except over the needles. He is intractable, and cries to go home; tongue clean; eats well. 24*th*.—After being fomented he passed his water freely and voluntarily, and finding it did not hurt him, has done so ever since. Appetite good; tongue clean. *March* 2*nd*.—No pain or irritation from the pins; dressings changed, dry lint only being employed; a drop or two of serum oozed from the punctures; no swelling whatever; eats and sleeps well; bowels opened by an enema. 8*th*.— A thick line of consolidation now felt in the track of the pins along the canal. Pins withdrawn with great ease; dress with dry lint. *15th.*—Punctures quite healed; the canal between them blocked up by a line of induration, completely obscuring the superficial ring. On coughing, a little more bulging is apparent on the side operated on than on the other, above the site of the upper cicatrix. Discharged; to have a light truss, worn only during the day. The mother promised to let me know if there was any return in this case. I have not heard from her to that effect.

Case 28.—Samuel East, æt. 15, operated on in King's College Hospital, April 20th, 1861, for a right, direct, scrotal rupture, of six years' standing. The neck of the sac very wide and lax, admitting the fingers freely into the abdominal cavity; the usual operation by wire (page 109) was performed, both ends of the wire being passed across the sac. *21st.*—Had considerable pain last night; allayed by pil. sapon. co., gr. v. at bedtime; no pain to-day; tongue a little white; pulse 100; no sickness; no tenderness over the abdomen; no tympanitis; respiration abdominal and equal in both groins. In a day or two, the pulse returned to its normal standard; the tongue lost its white fur, and no abdominal symptoms whatever appeared. Bowels open by castor-oil. On the 29th, sleeps without sedative; wishes for meat and porter; discharge moderate, from lower opening chiefly; induration apparent in the canal; no tenderness above. Middle diet; half pint of porter. *May 2nd.*—Continues well; pressure applied to the wire loop by spica bandage; consolidation more evident. *3rd.*—Had a little more pain last night from the pressure; none this morning; no increase of discharge; a little swelling evident in the testis, but it is not tender. *9th.*—Swelling of testis entirely gone. *13th.*—Wire removed without difficulty, from the lower opening; very little discharge. *20th.*—Wounds entirely closed; much consolidation; no cough impulse whatever; superficial ring quite obliterated. Discharged. To have a light truss fitted. *June 12th.*—Seen by me to-day; induration almost entirely disappeared; no cough impulse whatever; testis a little enlarged, but otherwise quite normal. On August 24th he was shown in the theatre of the hospital. The groin was in a most satisfactory condition, a slight bulge above the upper cicatrix being the only evidence of the hernia which had existed; the superficial ring was completely blocked by adhesions; the testis had resumed its normal size, and gave no trouble nor pain. Promised to apply in case of a return.

Case 29.—John Belman, æt. 22, a sailor, discharged from H.M.S. on account of a rupture, which had existed ten months. It was direct, scrotal, large, and on the right side. Operated on in King's College Hospital, May 4th, 1861, by the ordinary wire operation (page 107), one end of the wire only being passed across the sac. The rupture was easily reducible; the groins and genital had the imperfect fœtal outline strongly marked. 5*th*.—Did not sleep much on account of the pain; tongue white; pulse 80; belly soft, not tender on pressure; no swelling of the testis or scrotum. 6*th*.—Slept better; testis and scrotum a little swelled; tongue white; pulse 80. 7*th*.—Complains of a slight griping pain in the belly; no tenderness on pressure, no tympanitis, nausea, or sickness; tongue white; pulse 80. Bandage removed; water dressing; fomentations; pil. sapon. co., gr. v. every four hours. 8*th*.—Pain entirely gone; feels much easier; testis more enlarged; effusion can be distinguished in the fundus of the sac; little discharge of thick pus. 9*th*.—Bowels freely opened by castor-oil; belly soft and pliant; scrotum œdematous, testis enlarged, but of normal feel. 10*th*.—Has again a little pain occasionally in the bowels, but the tongue is cleaner, and he wishes for a mutton chop. 11*th*.—Pain entirely gone; swelling of testis and scrotum much diminished; testis not tender; scarcely any discharge from the wounds. 13*th*.—Apply pressure by spica bandage; middle diet, and half pint of porter. 18*th*.—Doing very well; sac can be felt in the scrotum above the testis, hard and contracted. 20*th*.—Wires untwisted, and one withdrawn without trouble and very little pain, the other removed next day; slight serous discharge. 22*nd*.—Patient about the ward; mass of consolidation evident in the canal. 30*th*.—Wounds closed; discharged. This patient was seen by me on June 7th, July 18th, and August 10th. The indurated mass gradually diminished, and at length disappeared; some varicose thickening upon the cord was then apparent, a little enlargement of the testis remaining. Thinking this might result from the pressure of the truss he was wearing, which was not a very good one, and which chafed the skin, I directed him to leave it off altogether. On October 26th, I again saw him. He stated that three weeks after he left off the truss, he having worked in the interval as a foundry man, while at work, he felt the rupture give way. On examination, I found that it had returned to about half its former size into the upper part of the scrotum, with a decided cough impulse. The adhesions had not been able to resist the test of lifting the foundry hammer without truss. Stated that he intended to have the operation repeated when he has a little holiday.

Case 30.—John Carter, æt. 18, operated on June 8th, 1861, at King's College Hospital, for a right, oblique, scrotal rupture of seven months' duration. The operation was the usual one by wire.

9th.—Passed a comfortable night by the aid of the pill, but had pain in the groin for an hour or two; tongue slightly furred; pulse 86. *10th.*—Had a good night, but has more pain this morning; no sickness, tympanitis, or tenderness anywhere in the belly except near the groin operated on; tongue and pulse as in last report; fomentations. *11th.*—Pain almost entirely gone; testis rather swollen; some œdema of the scrotum; bandage removed for the first time; very little serous discharge, which passes freely through the puncture. *13th.*—No pain since last report; a small quantity of pus at the upper aperture; tongue clean; pulse 76; bowels not yet open; swelling of the scrotum more defined, and the testis can be distinguished from an elastic tumour placed above it, which evidently arises from an effusion into that part of the infolded sac which remains in the upper part of the scrotum. *15th.*—No pain whatever; bowels not moved, although two doses of castor-oil have been given; hardly any discharge; the upper loop of the wire twisted more closely down into the puncture; to have a full dose of house medicine, and an enema. *16th.*—Bowels moved freely by the medicine; feels relieved. *18th.*—Is quite well in health; the swelling in the upper part of the scrotum is very tense, and moves freely above the testis. To-day I made a puncture with a grooved needle into its centre, and evacuated an ounce and a half of a deepish yellow clear fluid, which was found, by the addition of nitric acid, to be highly albuminous; the cord could then be clearly distinguished behind the folded sac; the testicle has nearly recovered its normal size. *20th.*—Doing very well indeed; no pain; very little discharge, hardly half a drachm daily; good appetite; bowels regular. *24th.*—Wires removed to-day without difficulty by untwisting the lower ends, and drawing on the upper loop, without dividing it into two portions, as in former cases. A hard mass was felt along the course of the canal from the scrotal to the groin puncture. *26th.*—A little more discharge since the withdrawal of the wire, and a slight return of the effusion into the fundus of the sac. *27th.*—On examining the parts closely to-day, a hard mass is found blocking up the external ring, a fluctuating tumour below it, and then, in the bottom of the scrotum, the testicle tolerably distinct, and hardly at all enlarged or tender; the scrotal puncture nearly healed; discharge very scanty; to have an evaporating lotion applied constantly to the scrotum, and a dose of house medicine. *29th.*—Up to-day, and about the ward, with a spica bandage and compress on the groin; swelling above the testis much diminished; no pain on pressure; wounds filled with granulations, and hardly any discharge; much thickening about the inguinal canal; scrotum well braced up to the external ring. *July 3rd.*—To-day the patient left the hospital with the wounds quite healed, and wearing a light truss. *Aug. 24th.*—To-day the patient was exhibited in the theatre of the hospital; has worn the truss in the daytime only when at work; the external ring on the side operated can barely be distinguished by the finger; cough impulse arrested at the internal ring above the upper cicatrix. A slight bulging at the point is observable on both sides equally; feels the side operated on stronger than the other; will apply immediately he observes any change; has to wear a light double truss for a few months. On Jan. 20th, 1862, this patient was again seen; no change had occurred; superficial ring quite blocked up by adhesions; a slight bulging apparent at the site of the deep ring; scrotum well tucked up on that side. *June 21st.*—Shown to-day in the theatre of the hospital; no change; the superficial ring quite impervious; the bulging at the side of the deep ring still present, and evident to some extent on both sides.

Case 31.—Joseph Andrews, æt. 3, a pale, unhealthy-looking child, with a congenital hernia affecting both sides, that on the right side descending into the scrotum. Two trusses have been tried, but failed in keeping up the right side; the last caused so much swelling of the testis that it was necessary to discontinue it. Operated on at King's College Hospital, June 8th, 1861, on the left side first with a view of closing the patulous canal. The rectangular pins were employed, and applied in the manner given at page 131, and in Figs. 40 and 41. The case was a favourable one for testing the effect of the pins upon the canal, as applied on each side in the two different ways described in the text. After four days the pins were withdrawn, and a compress and spica bandage applied. The child suffered very little from the operation, although in a sickly state of health, and with occasional cough; the right testicle was also at this time enlarged. Cod-liver oil and the syr. ferri iod. were at the same time administered. After a short time, however, diarrhœa, from teething, came on, and the oil was discontinued, as it appeared to sicken the child and to act on the bowels. Notwithstanding these unfavourable circumstances the pin punctures speedily healed, only a little serous discharge having been induced, and the closure of the ring and canal was very complete and satisfactory on this side. On the 11th July, 1861, the right side was operated on; the testicle had, by this time, somewhat diminished in size, and the right hernia had increased rapidly under the effects of the cough. In the second operation a more horizontal direction was given to the pins, as described at page 135 (Figs. 43 and 44). The child was taken home by the mother. On visiting him in the evening, he was found in a quiet sleep (without sedative), having been apparently as well as usual since the operation. 12*th.*—Has not cried much, nor appeared to suffer from the presence of the pins; bowels open, and urine free; was again asleep at the time of the visit; is taking no medicine whatever. 13*th.*—To-day the bandage and lint were removed; no redness nor discharge about the punctures; testis not more enlarged; no swelling in the groin; bowels open every day once; bandage and lint replaced. *July* 16*th.*—Pins removed to-day, as a considerable amount of induration was apparent in the canal, and slight œdema of the scrotum, and oozing of serum from the inner puncture. Has not cried more than usual since the last report, and appears to suffer very little; pad and spica bandage. 17*th.*—A good deal of consolidation apparent in the canal and scrotum; a slight serous discharge from the upper puncture, and a more purulent one from the lower; œdema disappeared; a light truss applied to the interval between the punctures. 20*th.*—Punctures entirely healed; no impulse felt on crying, nor increase of swelling in the groin; child thin and pale. The room in which it lives is close and confined, and the odour unpleasant; is in very unfavourable conditions for health; continue constant wearing of truss. 27*th.*—Child suffering much from diarrhœa and teething; belly tumid and enlarged, with the appearance and hard feel of mesenteric disease; an enlarged gland under the left ear, which is tender on pressure, and red. Ordered decoction of logwood mixture after each evacuation, and a powder of compound chalk powder and hyd. c. cret. night and morning. From this time to Sept. 21st the poor child was constantly ill from teething, cough, or diarrhœa; it was wasted and pale, with a tumid belly and enlarged cervical and submaxillary glands; cries and coughs a great deal; the groins continue, however, to resist these unfavourable circumstances. Saw the patient in the above interval five or six times; the left groin was uniformly close, firm, and resistant; the right testicle continued a little enlarged, but varied in size with the condition of the child's health, as did also the cervical glands. The empty sac of the hernia could be felt in front of the cord at the upper part of the scrotum, and seemed to be gradually contracting and shrinking upwards. On carefully examining the child during a paroxysm of crying, I could not detect any increase or swelling in the sac. The child was very thin and ill at the time with cough. *Dec.* 6*th.*—On examining the child to-day, I found him much improved in health and appearance, much fatter and stouter. The testis has now returned to its normal size, and the cervical glands nearly recovered. The mother states that two or three times, when the child was crying or coughing, while standing, and without the truss, she has observed a small protrusion below and to the outer side of the former rupture, but not one-third its size, and not descending into the scrotum; she states, moreover, that it never appears when the truss is on, as it used to do before the operation. Although the child cried much during my examination of the parts, I could not detect any protrusion; the sac could be felt, and appeared to have shrunken in size since my last examination. This case on this side, then, may be considered as not completely satisfactory at the present time; but it is reasonable to expect that the pressure of a proper truss will complete the closure of the internal opening, as the operation has done to the external part of the sac after a little time has elapsed. *March* 16*th*, 1863.—To-day I saw the child again; his mother informed me that there had been no return of the protrusion on the right side since I had last seen him, however much the child cried. There was no protrusion felt, on coughing without truss; the left groin remained perfectly cured.

Case 32.—Richard Henry Terry, æt. 5, operated on at King's College Hospital, July 20th, 1861, for a left congenital hernia, descending into the scrotum freely. The operation was done by the rectangular pins, as in Figs. 40 and 41. During the whole course of the case not a single symptom worth recording occurred; the child not suffering in the least from the presence of the pins. The spica bandage was removed three days after the operation, and the lint five days after. At this time there was no irritation whatever of the puncture, and throughout the whole case only a little serous discharge. The pins were removed without difficulty, and with little pain, on the 27th of July, when a truss was immediately applied between the punctures, and the latter dressed with dry lint. In six days more the punctures were quite healed, and the patient left the hospital with instructions to wear the truss constantly for two or three months, and to apply again if any return was observed.

Case 33.—Charles Vellam, æt. 13, operated on at King's College Hospital, Aug. 17th, 1861, for a left, oblique, scrotal hernia, of one year's standing. In this case the larger rectangular pins were applied longitudinally. From the straitness of the canal some doubt was entertained whether the invaginating finger (the little finger was used) was passed fairly behind the internal oblique muscle; and consequently, whether the conjoined tendon was taken up in a satisfactory manner. 19*th*.—No symptoms; no pain; bandage removed. 21*st*.—Lint and straps removed; no suppuration. 24*th*.—Pins removed easily; a drop or two of pus followed their withdrawal. Has had no pain whatever except when the parts were touched; bowels open three days ago by a little castor-oil; to have middle diet, and half-a-pint of porter. *29th.*—Continues in good health; upper opening nearly healed; lower admits the probe to the depth of three-fourths of an inch, and discharges a few drops of thin pus daily; a hard cylindrical consolidation evident in the track of the pins completely obscuring the external ring. A little œdema of the scrotum, and slight enlargement of the testis is apparent, with some tenderness; testis to be supported and fomented. *Sept. 1st.*—Doing very well; testis still a little tender, but not larger; lower puncture still discharging a drop or two of pus. *5th.*—Lower puncture healed; tenderness of testis diminished. *7th.*—Tenderness gone; upper opening still oozes serum, although a very minute point. *9th.*—Both punctures healed. *11th.*—To-day the patient left the hospital, wearing a truss fitting between the punctures. *Sept. 21st.*—Exhibited to-day in the theatre of King's College Hospital. The groin is firm and resisting; a slight bulging apparent at the upper puncture when he coughs; the pressure of the truss has caused a blister over the pubic spine; the truss was directed to be altered; but this being neglected, on Oct. 21st he applied at the hospital with a swelling of the testis and enlargement of the cord on the side operated on. The swelling is tender on pressure, and gives the feel of varicose veins; no impulse on coughing whatever. To take a dose of opening medicine, and to leave off the truss altogether, and foment the parts. On the 26th, the tumour was less tense, and could be felt to fluctuate from fluid contained in the fundus of the sac. No impulse is felt on coughing by the finger placed on the external ring; testicle less tender, and more of its normal size; fitted with a light truss and soft pad. *Nov. 24th.*—Brought to-day to my house. I find the truss is so weak now as to be powerless. There is now very little enlargement of the testis, and the wormy feel of the cord has disappeared; there is, however, more impulse in the canal than on my last examination; the bulging at the upper cicatrix, which always existed, is now more directed in the axis of the canal. Probably, in this case the conjoined tendon was not taken up by the pin; and the effusion in the sac and enlarged spermatic veins, irritated by the galling of the badly-fitting truss, have contributed to open up the hernial canal to some extent; truss fastened so as to press more upon the internal ring.

Case 34.—Robert Headlin, æt. 5, operated on at King's College Hospital, Aug. 24th, for a right, oblique, inguinal hernia, of six months' standing. The external ring admitted easily the little finger, and the hernia descended freely into the scrotum upon the testis. The smaller kind of pins were used in this case, and turned upon each other two turns, so as completely to twist up the sac. 27th.—Has had no symptoms and no pain whatever; bowels not opened; bandage removed. 29th.—Dressings removed; no pus; slight chafing by the point of the pin on the scrotum. Dressed with oiled lint; no bandage. *Sept.* 3rd.—Pins extracted; a hard cord apparent in their track; no suppuration whatever; no swelling or tenderness of the testis or scrotum. 5th.—Doing extremely well; no pain whatever. 7th.—Punctures very nearly healed. To have a soft light truss fitted between the punctures, and to get up and walk about. 11th.—To-day the patient was discharged with the punctures healed; a hard, firm groin; no cough impulse; less impression on crying at the groin operated on than on the opposite; consolidated band obscuring the external ring. To be brought back to the hospital directly any change is observed.

Case 35.—Thomas Wheatley, æt. 57 (looking much older), operated on at King's College Hospital, Aug. 24th, 1861, by the usual wire operation, for the cure of a left oblique hernia, descending half way into the scrotum, with a wide, loose, and patulous neck of the same diameter as the fundus. It had been observed three months only, but was rapidly increasing. Both groins very lax and bulgy. The patient had a nervous, irritable, sallow look, with a yellowish tinge of skin. For a week or two past has had a bad diarrhœa, which has weakened him much. Was very anxious to have the operation done at once. Complained, after the operation, of pain, and also of the tightness of the bandage, which was thereupon slackened. Milk diet; pil. sapon. co., gr. v. every six hours. 26th.—Has been very restless; complains of pain in the groin; respiration abdominal; no tenderness nor tympanitis in the belly; pulse 72; tongue covered with a white fur; fomentations over the bandage. 27th.—Much easier to-day and yesterday; no sickness; takes his food with relish, and feels hungry; pain on pressure over the groin operated on; respiration abdominal; tongue white; pulse 72; some œdema of scrotum; bandage removed; scrotum supported by straps of adhesive plaster across the thighs. Continue the pills; hst. ammon. acet. ë. sp. æther. chloric., three times a-day. 29th.—Continues much easier; tongue still furred, drier, and slightly brown in the centre; pulse 70, compressible; bowels not open since the operation; dressings removed; considerable discharge of creamy pus; skin between the punctures red and swollen, the redness extending upwards along the groin. At the lower aperture a slough of the fascia is apparent, and the odour is slightly fœtid; the upper puncture is matted up round the wire by inspissated pus. Since the removal of the bandage, the patient has moved about much in the bed, and the loop of wire seems to have been caught in the bed-clothes frequently; wire unhooked, upper end untwisted and cut off short; dressed with strips of wet lint and oilskin. The redness and œdema of the scrotum nearly gone; testis slightly tender. To discontinue the opium, and take acid. sulph. dil. ℞ x., sp. æth. chloric. ℞ x., and decoct. cinch. ℨj. t. d. s. *30th.*—Feels much better to-day, but complained of seeing spectra during the night, and brooded much over some depressing family matters. The night nurse reports that his mind wandered a little in the night. He looks rather excited; tongue less furred; pulse 68, compressible. On ordering 6 oz. of wine daily, I find that he has been having every day since the operation 6 oz. of brandy without my knowledge; it being my rule to administer no stimulants whatever in these cases unless the symptoms decidedly require it. To have a chop if he can eat it, and continue the brandy, but no wine. *31st.*—Last night the delirium returned much more severely; he became violent, got up out of bed, tore off all the dressings, and tried to get out the wire, shouting and gesticulating, and behaving exactly like a man in delirium tremens. One drachm of laudanum, ordered by the house surgeon, was only partially effective in diminishing the excitement; pulse 100; tongue furred; respiration abdominal; no tympanitis whatever, nor tenderness on pressure over the belly and groin. Little discharge from the punctures, which show within a sloughy margin, and give out a distinctly fœtid odour. *Sept. 1st.*—Last night the symptoms were again very violent, much difficulty was experienced in keeping him in bed, and he was constantly fingering and tugging at the wire. He afterwards described the spectra he saw, as if the various articles of furniture in the ward were dancing about him, and faces leering at and mocking him. A drachm and a-half of tinct. opii was administered to him, which had the effect of throwing him into an uneasy slumber or stupor, in which state he continued at the visit in the middle of the day. The breathing was sonorous; the countenance dusky; the nails rather blue; free perspiration on the face; the pulse was 72, firmer and quite regular in its beats. The skin about the punctures was livid, and the discharge was scanty, serous, and fœtid. The wires were completely untwisted, but the attempt to withdraw them roused the patient from his partly comatose condition, and he resisted so much with both hands and feet, that it was deemed advisable to leave them in the wound for the present; water dressings and bandages. *2nd.*—The patient remained eight hours in the state above described. At night

CASE XXXV.

half a drachm more of laudanum was given, and the brandy increased to 10 oz. daily. He passed a quiet night, and in the morning felt much relieved, and said he had slept well, without dreams or spectra; pulse 74. Ordered to continue the brandy (10 oz. daily); and to take as much strong beef tea as he feels inclined for, with a mutton chop and eggs, and the following draught: quinæ disulph. gr. viii., morphiæ hydrochlor. gr. ¼, sp. æther. chlor. ♏x., acid. sulph. dil. ♏x., aq. ʒj., every four hours. *3rd.*—Much better to-day in every respect; tongue moist, less furred, not brown; is perfectly self-possessed; no more spectra or delirium; eats his food with relish; pulse 60, firmer; bowels open freely to-day, but not purged. No tenderness, tympanitis, or tumefaction whatever about the belly or groin; no sickness or vomiting; no swelling, tenderness, or œdema about the testis, cord, or scrotum. The punctures look much better; the sloughs are limited to the margins of the openings, and are becoming separated by granulations all round; they seem to have been produced chiefly by the constant friction and working at the wires in his delirium. The discharge is now plentiful, thick, and creamy, scarcely any fœtor being evident. The wires were to-day withdrawn, upward, very easily, bringing away a small film of slough from their track. *5th.*—Continues much better; quite free from delirium; eats and sleeps well; pulse 72. Bowels open to-day and yesterday; a long, thin thread of slough drawn out of the wire track; florid granulations freely springing up; discharge less profuse, thick, and creamy; continue water dressing, and diminish the brandy daily by 1 oz.; continue the quinine and morphia mixture. From this time he constantly and uniformly improved, sleeping and eating well, and looking better every day. On the 14th, he got up and walked about the ward, the openings being filled by healthy granulations, and ceasing to communicate with each other; pad and spica bandage. On the 18th he discontinued the medicine and brandy; and on the 23rd he was discharged. The punctures were very nearly closed, oozing out a little serum; a great deal of thickening in the canal; no impulse whatever on coughing. On Oct. 18th, the patient presented himself for examination, *not wearing a truss*, with a very favourable account of his increase of strength for work since he was operated on. The groin was found to be bulgy and fuller than the opposite one, but the cough impulse was arrested by the adhesions at the upper cicatrix; promised to let me know if the groin got weaker.

The delirium in the above case was clearly due to the continuous administration of brandy (6 oz. daily) to a man on low diet, unaccustomed to its use, reduced by a previous diarrhœa, and troubled (as he explained to me afterwards) by family matters. The symptoms were clearly those of delirium tremens, and were entirely allayed by the free use of opiates. The restlessness and exertions of the patient tried severely the immediate security of the sutures, but they resisted well, although not assisted by a bandage, and the case made a rapid recovery. Not a single abdominal symptom made its appearance in this case. I hear from Mr. Kingdon, surgeon to the City Truss Society, that this patient afterwards applied to him for a truss in Nov. 1862. Mr. Kingdon states that the rupture had again reached as far as the external ring, at which it protrudes. The patient stated that it reappeared six weeks after the operation. In this case the structural weakness originally apparent in the groin was further increased by the sloughing which followed the operation. Partly to this I attribute the ultimate unsatisfactory result, and in part to his not having worn a truss after the operation, to protect the newly-formed tissues during labour.

Case 36.—Mary Dobson, æt. 27, operated on in King's College Hospital, for a right, oblique, labial hernia, of four months' standing. The operation was a modification of the usual wire method without transfixion of the sac, but simply twisting up the wire after the inner and outer boundaries were secured in the loop. Operation performed Sept. 21st, 1861. *22nd.*—Very little pain; does not complain. *24th.*—Bandage removed. *28th.*—Going on extremely well; tongue clean; bowels open; no symptoms whatever. *Oct. 1st.*—Consolidation in the wire track evident; very little discharge; no pain. *10th.*—Wires removed to-day without difficulty. Water dressing: bandage. *14th.*—Got up to-day and about the ward; discharge slight and serous. *17th.*—Punctures quite healed; much thickening; is in excellent health. To have a truss fitted, and to be discharged. *Nov. 30th.*—To-day, the patient presented herself for examination, wearing a light truss. A hard cord blocks up the external ring, and extends quite up the canal. A little bulging at the internal ring; no cough impulse whatever. On the last occasion that I saw this patient, a few months ago, the bulging in the groin remained in the same condition, a slight impulse being also felt when the patient coughed violently without truss. Recommended to continue the use of the truss so long as the weakness remained.

Case 37.—John Smith, æt. 23, a prisoner in the Millbank Penitentiary. Had a large, direct, scrotal hernia, of two years' standing, on the left side, with a very lax internal opening, admitting three fingers. Operated on October 5th, 1861, by wire in the ordinary way; during the passage of the wire through the conjoined tendon it gave way and broke off; the needle was reapplied, but with less certainty as to the hold upon the conjoined tendon and edge of the rectus. Milk diet; pil. saponis co., gr. v. h. s. s. *7th.*—Doing extremely well; no pain in the belly, which is soft and compressible; tongue clean; pulse 72; skin open; eats and sleeps well. Bandage removed; very little serous discharge; leave off pills. *13th.*—Doing very well; has not had an unpleasant symptom; slice of meat, and wine ʒiv. daily. *20th.*—Removed the wire; very little discharge; induration not well marked; pad and spica bandage. *Nov. 4th.*—During the last week, after he had been up and about the ward for some days, while at the water-closet, straining hard, and without truss, he felt something give way in his groin, and the tumour reappeared; at first but slightly, and to the inner side of the scar; but afterwards on the outer side also, after another sense of something giving. The hernia, on examination, proved to be in the same condition as before the operation. The patient much wished the operation to be repeated. Accordingly, on the 28th of December, 1861, this patient was again operated on, by the wire operation figured in Fig. 36, page 123, crossing the wires within the canal over the deep opening, and using a broad boxwood compress to fix the wires at the surface. *30th.*—Patient doing very well; some swelling of the testis; not a single bad symptom. *Jan. 6th*, 1862.—Doing very well; very little discharge; testis enlarged, but not tender on pressure. *15th.*—Much induration in the canal and about the groin; wires untwisted and pad removed; some slight febrile symptoms; a small abscess, which had formed under the skin of the groin over Poupart's ligament, was opened, and about a drachm of thick, yellow pus evacuated. *23rd.*—Wires withdrawn with great ease; induration very extensive and hard; punctures filled with granulations; no cough impulse was apparent. From this time up to that of his discharge from the prison, I was informed that no return of the rupture had occurred. When he left, he wore a truss, and was recommended to apply to me if the tumour returned, which he promised to do. I have not seen him since that time.

Case 38.—James Perkins, æt. 18, operated on in King's College Hospital, for a left, direct scrotal hernia, of twelve months' duration; the finger passed freely into the abdominal cavity. The usual wire operation was performed, Oct. 19th. On the evening visit, he was found asleep. *20th.*—Tongue rather white; pulse 96; skin hot; no tenderness of the belly, but the testes rather tender, and the scrotum œdematous. A gonorrhœa has appeared since the operation; ardor urinæ and discharge. To syringe warm water into the urethra frequently during the day. Haust. effervescens ʒj 4tis horis sum.; pil. saponis co., gr. v., n. m. que. *21st.*—Better; œdema gone; no pain, hardness, or tenderness in belly; tongue cleaner; pulse 95; much scalding. *22nd.*—Bandage removed; testis still tender and enlarged; much gonorrhœal discharge; tongue cleaner; pulse 96; no sickness; belly soft; eats well, but does not sleep well. *23rd.*—A good deal of fever to-day; tongue white; pulse 100; has had pain across the back, but it is better to-day; abdomen carefully examined; no tenderness or distension felt on pressure, either deep-seated or otherwise, anywhere, except in the immediate neighbourhood of the wounds; testicle still swollen, and scrotum more œdematous; a small collection of matter had formed at the lower puncture (which has nearly healed by the adhesive process). This was evacuated and well pressed out; no sloughs or redness; the scrotum is puckered up to the lower opening and obstructs the exit of the discharge. Dressed with wet lint, and without bandage. To have castor-oil ʒss. *24th.*—Fever still continues; bowels not yet opened; testicle still tender, but the œdema of scrotum has nearly disappeared; gonorrhœal discharge profuse. *28th.*—Feels better; bowels freely opened twice by the oil; tongue less furred; pulse 94; skin less hot; more discharge from the lower puncture; very little from the upper; testis less tender and swollen; wishes for meat. To have 4 oz. of wine daily, and a slice of meat, if better, to-morrow. *26th.*—Very much better; pulse 92; tongue cleaner; more discharge from scrotal puncture, and also from urethra; less thirst and fever; complains only of scalding; eats his food with relish. *28th.*—Continues much better; fever gone; pulse 86; tongue slightly furred; skin cool; no tenderness of abdomen, which is quite soft; less discharge from the punctures, more from urethra. *29th.*—Improving daily; complains only of scalding. To have lotio plumbi injected five or six times a day, and nitric acid and cinchona mixture. *31st.*—Gonorrhœal discharge much better; swelling of testes nearly gone; induration very evident in the track of the wire. *Nov. 2nd.*—Removed the wire to-day; doing very well; very little discharge; gonorrhœa better; eats well. *8th.*—Gaining strength fast; no scalding. Got up to-day and about the ward. *14th.*—Punctures very nearly healed; to be fitted with a truss and discharged. As the gonorrhœa still continues, to have a copaiba mixture three times a day. *Dec. 2nd.*—Showed himself in the operating theatre to-day. No impulse whatever on coughing; external ring completely closed up; no enlargement of the testis; gonorrhœa well; to let me know directly any change occurs. *Feb. 2nd,* 1862.—Came to the hospital to-day, wearing a light truss. The parts are very firm; the superficial ring entirely closed; no impulse whatever; testicle normal in size and feel. Is much stronger in the groin since the operation. Will let me know if any change occurs.

Case 39.—Lewis Solomons, æt. 2½ years, operated on by pins for a large, congenital, scrotal hernia on the right side, Oct. 19th, 1861 He is a very fat child, with a violent, petted temper, and cries a good deal. 20*th*.—Pretty well to-day; has not cried much; bowels moved freely; eats well; abdomen soft; some œdema of scrotum. 21*st*.—More swelling of scrotum to-day; the end of the pin chafes the skin there, and it is sore and painful, the dressing having been wetted with urine and shuffled off. Bowels open; rather feverish. Poultice to the scrotum and lint to protect the skin from the pin. 22*nd*.—More restless last night; was sleeping when visited. Less redness and swelling of the scrotum; slight oozing of serum; eats well; bowels open. 23*rd*.—Feverish and fretful; skin covered with a red rash, and a cough has appeared; considerable œdema of the scrotum and prepuce; slight discharge of pus; passes his water well. On account of the appearance of the rash, and the prevalence of scarlatina in the neighbourhood, it was thought better to withdraw the pins to-day. Saline mixture and poultice. 24*th*.—Fever and rash continue; swelling of scrotum much less; a little discharge from the punctures; bowels open. 25*th*.—Much better; swelling nearly gone; very little discharge. *Nov.* 4*th*.—There is now much induration in the canal; no cough impulse whatever, though he cries and struggles a great deal when touched, using much exertion; punctures nearly healed; to have a truss fitted. 11*th*.—Came to the dispensary to-day with a light truss on. Groin in every way very satisfactory; scrotum shrunken in size; sac can be felt empty and shrivelled. *May* 19*th*, 1862.—To-day the child was shown in the operating theatre of King's College Hospital, and examined by numerous visitors. No tumour whatever has appeared since the operation. Child looks much better, and is stronger and more healthy than before. Still wears a light truss. On taking off the truss, and causing the child to stand up, no impulse whatever can be distinguished on coughing or crying; the superficial ring is well closed by adhesions; scrotum much contracted; testis normal. Has been seen since quite lately, and remains quite well.

Case 40.—John Bodal, æt. 25, an Irish bricklayer's labourer, applied at King's College Hospital with an enormous direct inguinal hernia, on the right side, which came suddenly after a bad fit of coughing, a fortnight ago. The rupture is of the size of two fists, protrudes readily, and swells enormously on coughing. Operated on by the ordinary wire operation (page 109) Jan 11th, 1862. The bandage was kept on till the 14th; the patient having slept well each night; very few or no symptoms being present. On that day complained of more pain in the groin; bandage removed. 15*th.*—Still complains of pain in the groin; none in the belly; no tympanitis; bowels not open; very little discharge; pulse 79. To have castor-oil. 16*th.*—Much more discharge; less pain; pulse 80; bowels freely opened; poultice applied. 23*rd.*—Has been doing very well; bowels open regularly; eats and sleeps well. A small superficial slough at the lower opening. 27*th.*—Slough removed; wounds look well; water dressing. 30*th.*—Much consolidation; testis a little swollen. 31*st.*—Wires removed without difficulty; canal filled by hard effusion. *Feb.* 10*th.*—The openings have now entirely healed; no impulse or protrusion whatever on coughing; the hernial opening seems to be quite closed. Discharged, wearing his old truss. *May* 7*th.*—Applied to-day; has been working ever since his discharge as a Covent Garden porter, lifting very heavy loads on his back. Has had latterly, in addition, a very bad cough, which has given him some pain in the groin. The superficial ring is now more evident to the finger, but the cough impulse does not seem to affect it. *June* 20*th.*—To-day he again applied, complaining that on taking off his truss the other night, after a hard day's work, and a severe coughing fit, a small tumour appeared in the groin. On examination the ring feels more patulous, but still the tumour does not at once appear on coughing. The sides of the ring are hard and resisting. Recommended to get a lighter employment, and to wear his truss night and day while he has the cough. I have since seen this patient several times; on one occasion a tumour of the size of an egg was evident when he coughed very hard. The protrusion had evidently returned to some extent, but was not a tithe of the size that it was before the operation, and is now easily retained by the truss. Says it has appeared three or four times to some extent, but never unless the truss was removed. Cannot get a lighter employment. This case, in consequence of the poor fellow's necessity of doing very heavy work for the support of his family, and his violent cough so soon after the operation, has thus returned to some extent. The directness of the opening into the abdomen, and, perhaps, the adoption of the simpler operation for its cure, although so large, have no doubt rendered the adhesions less efficient in retaining the rupture. The powerful muscular development in the patient, especially of the recti abdomines, has also aided in reproducing the rupture to the limited extent which it now possesses. Another operation was proposed, affording, as I believe, a very good chance of ultimate complete success, but was declined by the patient for the present, on the score of his necessity for daily provision for his family.

Case 41.—H. C., æt. 28, a surgeon in the Indian army, who has himself many times successfully operated on natives in India by my first method, applied to me for operation on an oblique inguinal hernia on the left side, of eight years' duration. Has been on service in India the last four years, and finds the rupture very inconvenient, often painful when riding, and getting larger, notwithstanding the various trusses by the best London makers which he has tried, and which sometimes fail to retain it. The trusses he has worn have all had convex pads. The rupture is now scrotal, and of the size of a duck's egg. The operation was performed Feb. 28, 1862, by the use of wire in the usual manner (page 109). Had some pain during the night, allayed by pil. sapon. co., gr. v. The day after, he had a difficulty in passing water, more from the painful effect of muscular contraction upon the groin than from obstruction. A catheter was passed, and the water drawn off. A slight enlargement of the left side of the prostate was observed. Next day, an elastic catheter was passed, after which it was no more needed. Some irritation, scalding, and discharge from the urethra followed, but subsided in a few days. Five days after, some flatulent distension of the abdomen was present, but no tenderness on pressure anywhere but in the groin. No sickness or nausea. A dose of castor-oil opened the bowels freely, and removed all uneasiness. Takes his food well, with a glass of sherry daily. The testis began to swell a little two days after the operation, remaining larger and a little tender for three weeks, then resuming its natural size and feel. Moderate suppuration was present after the fifth day. The wires were removed easily and together, March 21st (three weeks after the operation). On the 27th, the apertures were so far healed that he got up, wearing a light truss of the horse-shoe pattern. An injection of ten grains of tannin to the ounce of water was used, for a week before the final closure of the punctures. On April 12th, both wounds were entirely healed. On May 5th, the scrotum was much tucked up, and the superficial ring closed by a hard cicatrix. A very slight bulge was apparent, on coughing hard, at the site of the internal ring on both sides; but not more on that operated on than on the other. Feels the side as strong as the opposite one. The irritation of the urethra has entirely disappeared. *June* 10*th*.—Parts remain unaltered, even during the hard course of coughing the patient was put under. On July 1st, a slight giving was observable at the site of the conjoined tendon, opposite the superficial ring, when the patient made a violent voluntary compression of the recti, holding his breath at the same time. No bulging whatever was, however, then felt at the internal ring; no cough impulse whatever; feels very well and strong; no return of urinary symptoms. *Sept.* 10*th*, 1862.—Nearly seven months after the operation, he called on me previously to his departure for India. Has been married since the last date, and travelled over Switzerland. Looks well, and feels in first-rate health, and much stronger than before the operation. The side operated on has not changed at all; no weakness, bulging, nor impulse on coughing whatever, is felt during the examination, although making considerable muscular efforts. The conjoined tendon seems to have recovered its tone and resisting power perfectly. Close above the pubis, at the inner side of the cord, a very small opening in the superficial ring is apparent, not admitting the point of the finger, and giving no impulse on coughing. Above, this is circumscribed by a firm, strong band of adhesion, uniting the pillars of the ring. Testis and cord perfectly normal; no return of urinary symptoms; the patient is very much satisfied with the result. Recommended to continue the horseshoe truss for some months longer.

Case 42.—William Stride, æt. 22, operated on March 29th, 1862, in King's College Hospital. Has a double oblique inguinal rupture; that on the left side has been observed five years; it is now scrotal, and of the size of an orange, with lax orifices and wide neck. On the right side is a bubonocele, lately appeared. Trusses inefficient in keeping up the left rupture at his work, in the government factory at Woolwich. Left side operated on by the usual wire suture. Slept the night after the operation with the aid of pil. sapon. co. No tympanitis, nausea, or swelling of the testis supervened. Bowels opened by castor-oil three days after the operation; suppuration very moderate. Wires removed on the fourteenth day. Walked about the ward three weeks after the operation. Discharged three days afterwards with the punctures completely healed, and wearing a double horse-shoe truss. On May the 9th, and again on the 19th, was examined: no change was observed. The general impulse of the abdominal walls on cough was very markedly arrested at the upper cicatrix, affording a strong contrast to the opposite groin, in which it descended to the superficial ring. The ring on the side operated on is completely filled by a firm cicatrix. Promises to apply immediately any change occurs. *Sept.* 29*th.*—Saw this patient to-day; no impulse whatever on the side operated on; complete closure of the canal above the superficial ring. Still wears a truss during the day. Feels stronger on the side operated on than the other, which is unaltered in its appearance. Has since been shown in the theatre of King's College Hospital several times. In February, 1863, this patient was again examined; no change whatever was apparent on the side operated on; the external ring quite obliterated; cough impulse arrested at the deep ring; tested freely by coughing without truss. Still wears a double truss when doing work, to support the opposite side, which feels now much weaker than the side operated on.

Case 43.—W. B——, æt. 32, a gentleman of exceptionably nervous and susceptible temperament, a patient of my friend Dr. H. Stavely King, of Brook-street, applied with an oblique scrotal rupture on the right side, of the size of a large orange, which had been present for the last three years, and was very difficult to manage with a truss. Has worn trusses of every kind, all with convex pads. States that the rupture, and the opening through which it passes, have been steadily increasing in size, so that, at present, it passes directly out of the abdomen. The truss at present worn (the most effective he has had) evidently presses into and dilates the superficial ring. A decided varicocele is coexistent in the cord of the ruptured side, and can be felt with the finger to extend upwards into the canal. States that this increases enormously in size when straining at stool, even though the truss keeps up the rupture. The sides of the opening are slender and loose, and dilate easily under pressure. Much pain is experienced in invaginating the scrotum upon the finger into the canal. Is not in very good health, having had oxalates in the urine several times, and latterly been troubled with boils; is very anxious to have something done to cure the hernia. Operation performed by wire in the usual way, March 12th, 1862. The presence of the varicose veins in the canal prevented so high an application of the sutures as might have been desirable, the bunch of veins doubled up with the invaginated fascia into the canal, obscuring considerably the internal opening and edge of the conjoined tendon. The care required to avoid them with the needle contributed to prevent a distinct perception of the transfixing of the conjoined tendon. 13*th*.—Had hiccough and retching during the night; some tympanitis in the abdomen; pain at the site of the operation only; pulse 75; bandage removed; water dressing; hot fomentations. 14*th*.—Fever apparent to-day; borborygmi heard; pulse 75. 15*th*.—Suppuration established; tongue white and furred; pulse 75. 16*th*.—Complains much of flatulence; enema of oil of rue ʒj. and gruel, which brought away no results. 17*th*.—Had calomel and colocynth pill and black draught, which results in a large motion of a very offensive character. 18*th*.—Had a disturbed night; flatulent distension entirely disappeared; pulse 90. Dressings at the lower wound discoloured by dark-coloured blood; suppuration at upper wound favourable; water dressing. 19*th*.—Better night; pulse 75; tongue cleaning; slight oozing of blood at the lower puncture continues; a few granulations apparent at the punctures; complains of pain in the testis, which is slightly enlarged. Wound syringed with tepid water. 20*th*.—Much disturbed by incessant visits of friends; restless; pulse 95. On endeavouring to pass a stool, more venous bleeding appeared at the wound. Repeated the pill and black draught, which opened the bowels very freely; red lotion to the wound; to have the wine increased. 22*nd*.—Reappearance of sanguineous stain in the dressings; painful and spasmodic contraction of the cremaster muscle, especially at stool, which results in an increase of the venous bleeding; this was allayed by morphia. Punctures injected with a solution of tannin, and a pledget of lint dipped in it applied. To have tinct. ferri muriatis, with quinine; tongue cleaning steadily; sleeps well with morphia. 23*rd*.—A small, black, hæmorrhagic point appeared in the lower wound upon the cord; upon this a few drops of tinct. ferri muriatis were dropped. Much disturbed by the visits of friends, in spite of the emphatic warnings of Dr. King and myself; pulse got up to 90. Continue iron, quinine, and wine. 24*th*.—Is better; pulse 85; slight staining of the dressings still present; bowels not open since last medicine. A warm aperient prescribed, with the effect of a copious action of the bowels. 25*th*.—Bowels acted again; complains of much pain in the rectum on voiding the motion, which was found, on examination, to be caused by hæmorrhoids; shivered slightly a few times; pulse 90 at night. 26*th*.—The communication of some news, which could not be kept from him, and which led to an interview of some hours' duration with a friend, in defiance of all remonstrance, has had the effect of depressing the patient very much; pulse rises at night from 90 to 110. Complains of much pain in the rectum; had an enema of morphia and oil, which gave him great relief; slept for three hours; on waking, his pulse was 85, at which it continued all day. To increase the stimulus. 27*th*.—A bandage applied over the wire loop, which, causing much pain, was removed after four hours. 28*th*.—Spasm in the rectum continues at intervals, and flatulent eructations during the day. Impossible to keep him quiet from visitors; after such interviews the pulse rises to 100; respirations 25. 29*th*.—Wire removed tolerably easily; very slight oozing of blood afterwards. Wounds have not at all a favourable appearance; a great deficiency of action is observable; discharge freely thin pus; pulse 120. Brandy and quinine given freely. Slept during the day under the influence of solution of morphia, after which the pulse was 98, and at night fell gradually to 85. During the night an increase of hæmorrhage occurred, following and induced by very severe spasmodic pains in the rectum, with fearful straining and ineffectual efforts at stool. Ice applied assiduously to the wound. Enemata of olive-oil and gruel bring away dark scybalous masses, and a full dose of castor-oil, with henbane, afterwards produces copious dark and offensive stools. The tenesmus continuing, a suppository of morphia is applied, and the abdomen rubbed with oil and morphia; passes a great deal of flatus; slept well during the night.

31*st*.—On firm pressure being applied to the sides of the upper puncture, a considerable amount of thick, inspissated pus, mixed with coagula, is pressed out. The nervous irritability of the patient has hitherto prevented any pressure on the wound. A great want of action is apparent in the wound. After this, no more oozing of blood occurred, but the discharge was thin and unhealthy. The patient is evidently labouring under some depressing influence, which seems to reside in the close and confined nature of the room in which he lies, and the smell of the new paint, which at this time pervades the house. The tongue is dry and brown; the pulse frequent; the skin muddy; the patient extremely irritable. It is accordingly determined to remove him to the house of his friend, Dr. King. This was done in one of Alderman's invalid couches, with a marked and immediate improvement in all the symptoms. The wound assumed a much more favourable appearance, and filled up gradually, but very slowly, the tannin injection continuing to be used with a view of consolidating the newly formed adhesions. A slight return of the spasm from the hæmorrhoids was successfully subdued by ung. gallæ co. Under quinine and generous diet he gradually recovered; and on the 5th of April a spica bandage and pad was applied, and he got up to the sofa. On the 12th, he went to Brighton, and in about a fortnight afterwards the punctures were entirely healed. The patient wore a horse-shoe truss-pad during the day. He has been under observation constantly since, and the parts appeared firm and resisting, until about three months afterwards. About this time the patient had increased much in corpulence, and was attacked by a violent cough, which affected him chiefly at nights, and came on in violent fits while the truss was not worn. After this time, a certain degree of weakness was observed at the superficial ring, indicating a shifting of the adhesions of the invaginated fascia close to the cord. The sloping rotundity of the conformation of the groin renders the precise application of the truss-pad difficult to secure; and the yielding during a violent cough-impulse has gradually increased, so that at the present time there is, during coughing, a small protrusion of a doughy feel, evidently composed partly of indurated sac, and partly of fat. This, however, is very small compared with the original rupture, and is easily controlled by the truss. Dec. 3rd, 1862.—To-day Mr. B—— came to me for examination, having worn a lever-spring truss of the pattern given in Fig. 48. For the last two months the protrusion has seldom occurred. On examination, it is found that the finger cannot be made to pass the external ring, but there is some impulse felt in that position on coughing. The scrotum has lost its swollen and rounded appearance, but a small doughy tumour can be felt on coughing. Testis of the normal size; feels strong and well, and wears the truss with comfort and without under strap. In the foregoing case, the state of health of the patient, his extreme nervous susceptibility, the peculiarities of conformation, and the want of vigorous reparative action, all combined to prevent the issue being as complete and satisfactory as might have been desired.

Case 44.—William Rose, æt. 19, engineer at Woolwich Dockyard, operated on in King's College Hospital, May 31st, 1862, for a congenital scrotal rupture, reaching to the size of a duck's egg. Has worn a truss from the age of fourteen years, which has been effectual in keeping up the rupture. The superficial and deep rings were very patulous, admitting easily of invagination. No symptoms whatever supervened after the operation, which was done with wire in the usual way. The bowels were opened by castor-oil four days after the operation. Afterwards middle diet and one pint of porter; water dressing and spica bandage over the wire loop. *June* 11th.—Wire removed without difficulty. 21st.—Allowed to get up and move about the ward with bandage and pad. *July* 10th.—Wounds all healed; discharged with a firm closure of the rings; no impulse or bulging; wearing a light horse-shoe truss. Since this time he has been shown once or twice to the class in the theatre of the hospital, no change having occurred in the groin.

Case 45.—Henry Marsh, æt. 19, working in Woolwich Dockyard, admitted into King's College Hospital, June 18th, 1862, for an oblique inguinal rupture on the right side, descending into the scrotum. Noticed two years and a half ago when lifting a heavy anvil. Since then he has worn a double truss, as the left side is also weak and bulgy. On removing the truss, the rupture drops at once into the scrotum. The testicle on the ruptured side has not descended lower than the upper half of the scrotum, and is atrophied so much as to be only one-third the size of its fellow. The cord is short, and its veins very varicose; the rings are large, with lax yielding sides; conjoined tendon very evident; scrotum wide and flaccid. Operation done June 21st, 1862, in the usual manner with wire. In the night he had some nausea, and slept little; complained next day of pain in the abdomen, which, however, was quite soft, and not tender on pressure. These symptoms were entirely removed by a few doses of pil. sapon. co., with hot fomentations, after the pulse had risen to 108, with a slightly furred tongue. On the 23rd, the bandage was removed; and on the 25th, the bowels were opened by castor-oil. He then slept and ate well, and there followed a pretty free suppuration and much consolidation. On the 30th, the atrophied testis was enlarged to above the size of its fellow, with some œdema of the scrotum. On July 7th, half the wire was removed with some difficulty, on account of a kink which had become entangled in the tissues; wounds filling with granulations; discharge thin; eats and sleeps well, and looks better in the face. 14*th*.—Remaining half of the wire removed without difficulty; swelling of scrotum disappeared; testis about the size of the opposite one. 16*th*.—Wounds nearly closed; discharge small in quantity, and thicker; pad and spica bandage; to get up and move about the ward. 18*th*.—A small tender spot of redness, surrounded by a hard margin, apparent about three inches above the pubis, the hardness being continuous with the consolidation in the inguinal canal. A small opening being made, evacuated a slight quantity of pus; poultice. 21*st*.—All redness and tenderness gone; punctures nearly closed. On the 8th of August, he was discharged, having waited nearly a week for his double horse-shoe truss. The groin operated on was very hard, firm, and resistant. *Aug.* 30*th*.—Shown to the class in the theatre; no impulse, but a little bulging at the deep ring; superficial ring firmly closed by bands of adhesion; testis at present very little smaller than the opposite one; very little remains of the varicocele. Feels much stronger, and more able to work than before the operation.

Case 46.—Charles Trotman, æt. 18, boiler-maker at Woolwich, operated on in King's College Hospital, June 28th, 1862, for a right, oblique, scrotal rupture, of the size of an egg, originating in a blow on the abdomen seven years ago, for which he was rejected by a Government medical inspector. Three years ago, his truss was broken by the fall of some iron plates upon him, the rupture came down and was returned with difficulty. Since that time, no truss has perfectly retained it, and it is getting rapidly larger. Superficial ring very wide and patulous; conjoined tendon very easily distinguished. After the operation, the symptoms were so slight as not to merit notice. There was a slight enlargement of the testis. *July* 2*nd*.—Discharge established moderate; bowels opened by castor-oil. 3*rd*.—Nettle-rash on the arms. 7*th*.—Doing uncommonly well; eruption disappeared; discharge moderate; induration filling the inguinal canal. 14*th.*—Wires removed easily. 30*th.*—Upper aperture quite healed; lower nearly closed. Discharged cured, wearing a horse-shoe truss. *August* 30*th.*—Shown to-day to the class. The rings are firmly closed; much of the induration has disappeared; no bulging or impulse whatever on coughing; feels and looks very well; is stronger than before the operation.

Case 47.—George Dunn, æt. 7 years, son of G. Dunn, of West Bromwich, Staffordshire, was sent to me for operation by Mr. Lloyd, surgeon, for an enormous congenital rupture, on the left side, distending the scrotum to the size of a large fist, and reaching halfway down the thigh, nearly obliterating the penis by distension. The deep opening at the neck of the sac was exactly opposite the superficial ring, so that the rupture emerged close above the pubis at the edge of the rectus muscle, as in a direct hernia, the conjoined tendon being altogether wanting. The hernial apertures were so wide and patulous as to admit easily the points of four fingers, and occupied the whole distance between the rectus muscle and Poupart's ligament. One finger could be carried into the abdominal cavity so as easily to feel the common iliac artery and aorta pulsating, and to reach as far upwards as the posterior surface of the umbilicus. The sides of the opening were very lax, loose, and dilatable. The bowels passed freely into the sac, and distended it enormously on the child making any effort. The sac was thickened and hard, but permitted the peristaltic motions of the bowel to be perceived at the surface. The rupture was easily and completely reducible, returning immediately the pressure was removed. When he winced under the pressure, the bowels emerged with a strong rush which rendered it impossible to keep them in the abdomen except by turning him upside down. This position was afterwards found to be absolutely necessary during the operation, and subsequent dressings, when his violent crying and struggles militated much against the maintenance of the closure of the hernial opening. No truss whatever had sufficed to keep up the bowels for the shortest time. The rest of the abdominal region was small and contracted, from the habitual extrusion of so large a quantity of its contents. The boy was in strong, robust health, and well nourished. This formidable case, appeared, at first, almost a hopeless one, from its size and that of the hernial opening, and the great deficiency of the abdominal walls. At the earnest desire of the parents, however, I was induced to try the effect of an operation, more with the object of diminishing the size of the opening so much as to enable a truss to be worn with benefit, than with the hope of effecting a complete cure. On June 14th, 1862, in King's College Hospital, an operation was done with a large pair of rectangular pins, the upper and inner one passed through the tissues close to the border of the rectus, but not including it, and the outer upwards through Poupart's ligament. The pins were twisted and bound firmly down by strips of lint and plaster. The laxity of the structures, and the distance of the pillars at the pubis was such, that an opening remained after the twisting of the pins just above the pubis, through which a knuckle of intestine escaped. This was returned after relaxing the pins a little, and a pad held firmly over it by a spica bandage sufficed afterwards to retain it, notwithstanding the violent struggles of the boy as the effect of the chloroform passed off. The patient was placed in bed with the shoulders low and the pelvis and legs raised upon pillows to relieve the strain upon the pins. With the exception of the evening of the day of the operation, when he cried a good deal, and a slight whitening of the tongue for a few days, the symptoms were so slight as not to be remarked; there was no nausea nor even disinclination for food; nor did he cry much after the first day. The bowels were opened naturally three days after, when the bandage was removed. The pins were then found to have sunk into the swollen scrotum, and produced a deep ulceration; though but little discharge, and that thick and healthy, was apparent at the punctures. The œdema of the scrotum had extended to the prepuce, which was much elongated, and affected with phymosis, so as to interfere with micturition. The end of the prepuce was removed, and its lining slit up with the scissors. The pins were at the same time withdrawn; much thickening and consolidation being apparent in the neck and fundus of the sac, obscuring the testis and cord. The violent struggles of the little fellow, when the pins were withdrawn, again protruded a knuckle of bowel close above the pubis. This was returned without difficulty, a stout pad and spica bandage over a simple dressing keeping it in its proper place. Much induration followed, filling up the whole anterior part of the sac, and producing a thickened and resisting condition of the borders of the hernial apertures. The punctures were both entirely healed by July 19th, when, on examination in the upright position, it was found that some protrusion still occurred through the aperture, close above the pubis, and that it passed down behind the thickened and twisted part of the sac into its posterior portion, so as to distend the perineum behind the testis and cord. At this date, the boy, being brought again under the influence of chloroform in the theatre, was placed upon the table with the pelvis raised very high. The posterior part of the scrotum behind the testis was then invaginated into the remaining hernial opening with the forefinger. A large, common suture needle, armed with strong wire, was then passed along the finger and made to transfix the attachment of Poupart's ligament to the pubic spine. It was then carried, with its concavity towards the finger, under the skin, and made to emerge near the root of the penis. The wire being drawn through, the needle was again entered at the last-made puncture, and made to take up the attachment of the internal pillar of the ring to the pubic symphysis. It was then brought round the finger as before, but on the opposite side of it, and made to emerge at the puncture first made. The wire being thus drawn through under the skin enclosed the remaining portion of the unobliterated sac in a circle, which included also the attachments of the pillars of the ring to the pubis. Some difficulty was experienced, on twisting the wire up tight, in keeping the bowel from protruding, as the boy just then began to struggle violently. This was safely overcome by keeping the finger plugged into the opening until it was embraced closely by the wire loop, and then cautiously withdrawing it, twisting the wire tight as this was done. A pad and spica passed across the perineum pressed the wire firmly down upon the pubis behind the testis. The latter with the cord and obliterated portion of the sac could be distinctly recognised during the operation. No symptoms whatever followed this operation: the boy ate his dinner next day as usual, and the bowels were moved regularly. The wire was kept in fourteen days, giving rise to very little suppuration, and much induration. It was then removed, without difficulty, by dividing the loop with a fine pair of cutting pliers passed through one of the punctures, and then drawing it out by the twisted part. Firm pressure was then continued by pads, sticking-plaster, and bandage, over simple dressing. By August 8th the openings were completely healed, and the boy got up and walked about the ward. On examination without bandage and in the erect posture, a small degree of protrusion was still observed in the neck of the sac behind the obliterated part when the boy coughed, but the sac in the scrotum was completely obliterated, and contracted into a hard mass. The protrusion could be felt to escape through a chink left between the edge of the rectus muscle and Poupart's ligament above the pubis. The little fellow was in capital health, eat very well, and somewhat fatter than when first admitted. He had suffered so little from the last operation which had proved so successful in its main object, that I conceived that attempt might properly be made to close the remaining opening. Accordingly, on August 13th, the boy being placed under chloroform in the same position as in the last operation, the scrotum was invaginated up to the aperture or chink, and a small curved hernia needle was carried through the scrotum without previous incision, and passed through the edge of the rectus muscle a little above its attachment to the pubis. The point of the needle being passed through the skin of the groin in the usual way, a stout wire was hooked on and drawn through into the scrotum. The needle was then disengaged and passed in the usual way through the hard cicatrix involving Poupart's ligament, and then through the skin aperture above. The opposite end of the wire was then hooked on and drawn into the scrotal puncture alongside the first. The two ends were then twisted together and drawn up tight by twisting the loop above in the usual manner. The rectus seemed to be fairly held by the suture and the chink obliterated. A pad and bandage completed the operation. The boy went on eating his food as usual, the bowels were regular, and not a symptom supervened. *Aug.* 26*th.*—Sufficient induration being apparent, with very little suppuration, the wire was this day removed. On the 29th, the apertures were nearly closed. *Sept.* 10*th.*—After wearing a stout compress and bandage up to this time in the recumbent posture, the boy was examined in the erect posture without bandage. On coughing, a small bulging is still apparent just above the pubis, but no dilating impulse is perceptible in the thickened and amalgamated tissues at the neck of the sac. The protrusion is apparently composed of a mass of agglomerated sac, containing some omentum and bowel. The chink which still remains is evidently due to the impossibility of drawing close together the sides of the hernial opening, so widely separated at their attachment to the pubis. An ovate truss pad, with a central perforation, adapted for direct hernia (Fig. 50), and composed of boxwood, fits accurately upon the interval between the rectus and Poupart's ligament, and effectually restrains any bulging in this situation under any amount of coughing or exertion. The testis and cord of the side operated on can now be perfectly distinguished, the former being a little enlarged. The boy to wear the truss and move about the ward for a week, to see the effect of the truss upon the parts when moving about. He is then to be discharged. 30*th.*—Discharged; the truss keeps up the small remaining rupture perfectly.

Case 48.—Charles Dunn, æt. four and a half years, the brother of the last patient, admitted at the same time into the hospital, with a congenital scrotal rupture of the right side, as large as a small fist, which no truss had ever retained. The coverings of the rupture are so thin that the peristaltic movements of the contained intestines are distinctly visible through them. It is easily and completely reducible. The abdominal openings are very large and lax, and placed directly opposite each other, so that, as in the case of his brother, it has all the appearance of a direct rupture, the neck being bounded internally by the rectus muscle, and externally by Poupart's ligament; no conjoined tendon being recognisable. Two fingers can be passed easily into the cavity of the abdomen, feeling distinctly the pulsation of the iliac vessels. On the 28th of June, 1862, the operation by the use of large rectangular pins was performed in the usual way, the inner one transfixing the tissues close upon the rectus muscle, and the outer one Poupart's ligament. The experience of the previous case led me in this instance to place a couple of smaller pins transversely across the sac behind the larger. The deeper of these was placed close to and on a level with the upper border of the pubis, avoiding the spermatic cord on the other side. The superficial one was placed a little higher, close behind the larger pair. All protrusion below and behind the vertical pair of pins was thus prevented. A pad and spica bandage applied pressure over all. *29th.*—In the night he has shuffled off the bandage and a great part of the dressings, disengaging the small pair of pins from each other. The ends of the pins have disappeared under the skin, and could not be replaced, owing to the violent struggles of the patient. The pins were left as found, and a pad firmly placed above them, the spica bandage being stitched well to prevent future displacement. Considerable œdema of the scrotum and enlargement of the testicle followed the operation, but very little suppuration. The boy, with the exception of a slightly filmy tongue, continued unaffected in health; eat and slept well. Bowels open naturally two days after. *July 1st.*—To-day the pins were removed, the child being held up by his feet during the dressing to prevent any strain upon the new adhesions during his violent struggles. A stout pad held on firmly by sticking-plaster and bandage. *7th.*—The sac can be felt filled up to the neck by solid effusion; apertures nearly closed: very little discharge: is in capital health. *11th.*—A diminution is now apparent in the size of the solid tumour about the neck of the sac. *16th.*—To-day the child stood up without bandage for examination. When he cried violently, a bulge was observable in the neck of the sac extending about three-fourths of an inch downwards, the solidity of the fundus continuing firm and resistant. The scrotum is much shrunk and thickened. The testis can be distinguished somewhat enlarged. *July 19th.*—To-day, the aperture which remains at the outer border of the rectus was closed up by a wire ligature carried horizontally across by a common suture needle. One end of the wire was passed close above the pubis and made to take up the insertion of

Poupart's ligament, and the edge of the rectus tendon at its insertion. The other end was passed under the skin across the neck of the sac a little higher up, including the outer and inner pillars and conjoined tendon. The ends were brought through the same skin puncture on each side, and twisted tightly together down into each puncture, the twists being then hooked on to each other over a cylindrical pad of lint (see Fig. 38, page 127). After the operation, the child went on as usual, very little suppuration ensuing. On the 30th, the indurative action being considerable, the wires were untwisted and withdrawn. The apertures healed up speedily, and on the 12th of August, when a careful examination was made, the punctures were completely healed, the parts much consolidated and well closed in, and no impulse or protrusion whatever was apparent when he cried. The testis had nearly recovered its normal size, the shrivelled and contracted sac could be distinguished above it, not in the slightest degree affected by his crying. An ovate, perforated, boxwood truss pad was fitted on, and the child was taken home into the country by his father, looking fatter, if anything, than when he came to the hospital. Since that time I have heard many times from his father. The spring of the truss being rather too strong, caused at first some swelling of the testis and scrotum, and some excoriation of the upper cicatrix. This having been altered, the child wore the truss with perfect ease. Not the smallest protrusion has since occurred. *Sept. 30th.*—To-day, the child was brought to town for inspection. The hernial aperture remains completely occluded and firmly resistant to the cough impulse; the sac can be felt in the scrotum degenerated into a cord-like mass. The testis is of normal size and shape. The truss he is wearing has a very weak spring, barely sufficient to render support. *March 13th*, 1863.—I heard to-day from the father (a very intelligent man) to say that both the boys remained in the same condition in every respect as when they left the hospital. No return whatever had occurred in the younger, and the slight protrusion of the older boy was perfectly and easily retained by the truss.

Case 49.—Richard Humphreys, æt. 30, a puddler at Nant-y-glo Ironworks, near Tredegar, Monmouthshire, sent to me by Mr. Jeffries, the surgeon to the works, for operation, with a double, oblique, inguinal rupture. On the left side is a large scrotal rupture of the size of the fist, with a large lax abdominal opening admitting easily three fingers, produced by violent coughing, seven years ago. On the right side is a bubonocele, just making its way through the superficial ring. Has not been able to wear a truss, from the left rupture slipping down behind it while at work. Both sides are easily and completely reducible. Operation performed, on the left side, in King's College Hospital, by the usual wire method, July 12th, 1862; conjoined tendon not very evident. With the exception of the loss of sleep for one or two nights, he suffered very little after the operation. 14*th*.—Scrotum slightly œdematous; testis a little enlarged; no nausea or tenderness on pressure over the belly; no tympanitis. 16*th*.—Complained of increased pain last night; feels uneasy; bowels not open; bandage removed; to have castor-oil. 18*th*.—Bowels freely opened; feels quite right again; eats middle diet. A spica bandage applied firmly over the wire loop to set up more action in the posterior wall of the canal. 20*th*.—Bandage caused more pain; discharge moderate and healthy. 22*nd*.—Has little pain now; eats, drinks, and sleeps well. *Aug.* 6*th*.—Wires removed without difficulty (twenty-four days after the operation). 26*th*.—Progressing very well; looks fat and well; wounds filled with red granulations; discharge almost entirely ceased. 27*th*.—The punctures have now entirely healed. The superficial ring is completely closed by firm adhesions; scrotum slightly tucked up; great induration in the inguinal canal; no bulging or impulse whatever; discharged, wearing a double horse-shoe truss with boxwood pad. Promises to let me know if any change occurs. This case was kept longer under treatment and pressure, with the wires unremoved, because of the want of certainty that the conjoined tendon had been included in the ligature to a sufficient extent. The result of the case so far is favourable to the practice in such cases. On the 29th October I received a note from Mr. Jeffries, of which the following is an extract:—" You will be pleased to know that he has quite recovered his health, and has gone to work; the cicatrix is hardly perceptible, and altogether it is as neat a cure as any one would wish to look at. He trusts that you will allow him to have the other side operated on next spring."

Case 50.—Colonel Slade, æt. 10, a healthy-looking boy, operated on at King's College Hospital, by the use of rectangular pins, in the usual way, August 1st, 1862, for an oblique inguinal rupture on the left side, reaching into the upper half of the scrotum. Observed after carrying a pail of water, three weeks before admission. It is probable that, in this case, the sac was not interfered with. After the operation the boy's health continued as usual. The bowels were opened naturally three days after, shortly after which he resumed middle diet. On August 9th, the pins were removed without difficulty, very little discharge being apparent. A hard line of induration apparent along the canal and track of the pins. On the 27th, the punctures were entirely closed, and on examination in the standing posture, no bulging or impulse whatever could be detected on coughing. The superficial ring was quite closed; testis and cord perfectly healthy and mobile. Discharged wearing a horse-shoe truss. Ten days afterwards, he was again examined after going about his work as usual. No change whatever was perceptible. Since heard of as remaining well.

Case 51.—John Day, æt. $1\frac{1}{2}$ years, operated on Aug. 28th, 1862, at King's College Hospital, for a congenital rupture of the right side, of very large size, totally uncontrollable by any truss which has been tried. The pressure of the truss he has worn had produced much excoriation, and latterly a large glandular abscess in the groin. Unhealthy-looking child, with a tumid belly. Has suffered much from diarrhœa and intestinal irritation from teething, on which account the operation has been from time to time postponed, the rupture meanwhile increasing rapidly in size. Two small rectangular pins were applied in the usual way, the inner one including the thin border of the rectus, which, from the great size of the deep opening, formed the inner boundary of the neck of the sac. On the evening of the day of operation the child had bilious vomiting, and cried a good deal. The vomiting continued at intervals during next day, when the scrotum being œdematous, and the testis swollen, towards evening the pins were removed, and a pad and spica bandage applied over the groin. At night the vomiting had ceased, and the child slept under the influence of ℞ x. of tinct. camph. co. Next day the scrotum was very much swollen, and a slight slough appeared at the lower puncture. A hard swelling filled up the whole of the inguinal canal, and a little pus escaped at both punctures. No impulse or dilatation was apparent when the child cried. Simple dressing and spica bandage. To have wine 2 oz. daily, and strong beef-tea. Two days after the operation the bowels were opened naturally and freely. *Sept. 5th.*—Free discharge from the punctures of healthy pus, very hard and large consolidation, occupying the canal and scrotum; testis a little enlarged; child improving rapidly in appearance. *9th.*—A small slough has come away from the lower puncture. Scarcely any discharge from punctures, which are rapidly closing. Testis diminishing in size; no impulse or protrusion whatever on crying; an enlarged gland in the opposite groin; and the tumid, hard belly showing a tendency to scrofula, the child is put on two eggs, and a teaspoonful of cod-liver oil, three times daily. *15th.*—Looking wonderfully better; punctures entirely closed. The consolidation diminishing in size, but increasing in firmness; testis nearly its natural size. Runs about the ward wearing a pad and bandage. To have a boxwood, horse-shoe truss fitted. From this time he was kept in the house to watch the progress of the absorption of the effusion in the canal. This diminished gradually, and at length disappeared in a hard cicatrix, which closed the superficial ring and canal. No impulse or protrusion whatever in coughing or crying succeeded. Up to the present date (March 16th, 1863), I have many times seen the child or its mother; no protrusion or impulse whatever has been seen, and the child is in much better health than before the operation. The symptoms observed after the operation, in this case, are due partly to the chloroform, and partly to the bad state of the child's health at the time. This was such that it was only the rapid increase of the rupture, and impossibility of restraining it by a truss, that led me to operate, especially in one so young as one year and a-half. This is the most tender age at which I had hitherto operated. Notwithstanding the short time that the pins were retained, the action set up was very abundant, and the case was surprising from the rapidity and soundness of the cure.

Case 52.—David Williams, æt. 34, operated on in King's College Hospital, Sept. 5th, 1862, for a direct scrotal hernia, of the right side, of two years' duration. The operation was done by the use of wire, as described at page 109. The conjoined tendon was secured close to the border of the rectus muscle. After the operation, the patient did not lose a single night's rest, though he took no sedative, so little pain had he. *Sept. 8th.*—Bowels opened freely by castor-oil. On account of the cleaning going on in the hospital, the ward is at present much overcrowded, and most of the sores in the room have assumed a sloughy appearance. In this patient a slough has formed at the lower opening. The testis is enlarged, and slightly painful on pressure; the bandage and pad removed for the first time. 10*th.*—Looks pale and anxious, but does not complain of pain anywhere; no sickness, nausea, tenderness, or tympanitis whatever; discharge from lower puncture thin, sanious, and fœtid. Has a cough, and states that the sputa are tinged with blood. This, it appears, he has occasionally been subject to. Much induration around the wires; slough separating a little; discharge thin, copious, and foul at the lower, but purulent and healthy at the upper puncture. Considerable swelling of testis, but little pain on pressure. Discontinue bandage, and apply poultice to groin and scrotum; pint of stout daily; quinine, sulphuric acid, and chloric æther mixture. 17*th.*—Several pieces of slough removed since last report. 22*nd.*—The whole of the remaining slough removed to-day, appearing to include a portion of the spermatic cord. Discharge from upper wound ceased. 26*th.*—On examination to-day, the testis is found to have diminished considerably in size, but still gives its peculiar sensation to the patient on pressure. *Oct. 6th.*—Upper wound entirely healed; truss with ovate pad applied. *9th.*—In consequence of irritation of the cicatrix and considerable swelling in the scrotum, the truss has been discontinued for the present. A hard, painful swelling is apparent in the site of the testicle. *10th.*—An abscess which had formed in the scrotum has burst; the swelling has entirely subsided after the escape of some matter. The suppuration has evidently taken place in the interior of the testicle, the size of which is now felt to be greatly diminished, and its normal sensation gone; the upper puncture is firmly cicatrized. Truss reapplied; to get up and move about the ward. *11th.*—Discharged to-day with a very small wound still remaining in the scrotum; no impulse whatever at the abdominal rings. *31st.*—Came to show himself to-day. A slight bulge in the upper part of the inguinal canal is the only evidence of a rupture having existed. No impulse at the external ring whatever on coughing. Testicle quite atrophied. Will let me know if any return of protrusion occurs.

Case 53.—Robert Stephens, æt. $2\frac{1}{2}$ years, operated on at King's College Hospital, Sept. 20th, 1862, for a direct scrotal rupture on the right side, of the size of a hen's egg. The deep hernial opening is placed close to the border of the rectus muscle, close above the pubis; observed first a few days after birth; a bubonocele also observable on the left side. Operated on by the use of three pins, two placed longitudinally, and one across, close to the pubis, and behind the other two. Not a symptom occurred after the operation. His appetite and bowels were regular; sat up and played on the bed the day after the operation. Pins removed on the eleventh day, not having been meddled with before. Much thickening in the canal; no suppuration whatever. *Oct. 8th.*—Discharged with the punctures healed, wearing an ovate truss pad. No impulse whatever when the child cries; much consolidation in the canal. Seen again a month after; the consolidation had then disappeared. No impulse or protrusion apparent on crying.

Case 54.—Charles E. B——, æt. 21, operated on for an oblique scrotal rupture of some years' standing, which no truss had been found to retain, on account of the size of the deep opening and the great obliquity of the pelvis. Operation done by wire in the usual manner, Oct. 7th, 1862. No symptoms of importance after the operation. Bowels opened by castor-oil five days after. Some swelling of the testis; discharge thick and scanty; bandage retained untouched for four days; wires removed sixteen days after; much induration. *31st.*—Both openings nearly closed. *Nov. 1st.*—Removed from his lodgings to

Lewisham, wearing a truss over the dressings; a slight return of the swelling of the testis afterwards. 19*th*.—Called to-day after staying awhile in the country; a hard mass remains in the site of the scrotal sac; testis still rather enlarged. *Dec.* 2*nd*.—Came again to see me, having suffered from a bad cough and rheumatism; thickening about the cord nearly all gone; testis nearly of its normal size; no impulse whatever on coughing. *Jan.* 14*th*, 1863.—Called to-day. The parts have resumed a normal condition, all swelling and hardness having disappeared. No impulse or bulging whatever; on the opposite side there is more impression made by a cough than on the side operated on. Recommended to wear a double truss for some time. *March* 4*th*.—Remains well.

Case 55.—Thomas H. Hewson, æt. 9, operated on by the use of pins for a right, oblique, scrotal rupture of the size of a pigeon's egg, at King's College Hospital; the result of a fall a fortnight before. Operation done Oct. 12th. No symptoms whatever followed, the patient eating and sleeping as usual. 22*nd*.—A slight discharge; testis a little enlarged. 25*th*.—Pins removed without pain; a thick mass of lymph fills up the canal. *Nov.* 3*rd*.—Ring-truss pad applied; moves about the ward. 5*th*.—Discharged, both punctures healed. No impulse whatever on coughing; no bulge at the site of the deep ring; superficial ring quite closed. *March* 7*th*.—Seen to-day; parts unaltered.

Case 56.—William Piner, æt. 48, labourer, from Hounslow, operated on by the usual wire method, at King's College Hospital, Oct. 18th, 1862, for a right, oblique, scrotal rupture, of the size of a large orange, for which no truss had been available; no symptoms whatever after the operation; wires removed sixteen days after; much consolidation; wounds healed Nov. 13th; discharged on the 17th, with the superficial ring closed, and no swelling or tenderness of the testis or cord. *Jan.* 2*nd*, 1863.—Came to-day to show himself, having been at labouring work in the interval. With a view of rendering the pad more comfortable he has fixed some tow and folds of linen on the truss-pad, which has displaced the pressure, leaving the canal entirely unsupported; the result has been a partial return of the protrusion behind the adhesions at the superficial ring. The truss-pad altered to bring its pressure more fairly upon the canal.

Case 57.—Joseph J. Adams, æt. 19, a thin, sickly lad, subject to epileptic fits, operated on in King's College Hospital for a right, oblique, scrotal rupture, of some years' standing, with very wide open rings. Operation done by wire, Nov. 1st, 1862; no symptoms followed the operation. A week after had an attack of scarlatina, with sore throat, and the usual eruption. The wounds suppurated freely, but did not seem much affected by the disease. Wires removed on the 17th. On Dec. 3rd, the lower wound was entirely healed, and the upper one nearly so; fitted with a horse-shoe truss-pad, and discharged. On the 14th, shown in the theatre; cough impulse firmly arrested at the deep opening. On Jan. 26th, again examined (and also on March 16th), having had several severe epileptic fits in the interval, in one of which he cut his face very severely. The inguinal canal and rings quite unchanged; a firm hard cicatrix in their site; no impulse or bulging. The truss fitting rather loosely, was properly adjusted.

Case 58.—John Weller, æt. 42, a labourer, a stout, healthy-looking man, sent to me by Mr. Hart, of Welbeck-street, with a double rupture. Both ruptures were oblique, and enormously large; that on the right side was nearly the size of a child's head, and of ten years' standing; that on the left somewhat smaller, and of two years' growth, but gave him much pain and uneasiness. No truss whatever would keep up either rupture, and being thereby almost totally incapacited, he was urgent that an operation should be done on the side that gave him most pain. Operation done by cross wires (Fig. 36, page 123) on Nov. 9th, 1862; wires fastened over a boxwood pad. Bandage removed on the 13th; pad removed on the 15th; no pain, sickness, or tympanitis. Appetite uninterrupted since the operation; was eating heartily the day after the operation; the tongue barely whitened. 24*th*.—More discharge of a healthy character; no pain whatever. A very large mass of consolidation fills up the hernial rings, and extends into the sac in the scrotum; some swelling of testis and cord; takes middle diet and a pint of stout daily. *Dec. 3rd.*—Swelling of testis and cord entirely disappeared. 7*th.*—Wires withdrawn without difficulty; very little discharge; a large amount of consolidation all over the groin. 10*th.*—Up and about the ward to-day. 17*th.*—Discharged with a double truss with ring pads, a large one on the right side, and a smaller on the side operated on. This is effective in retaining also the side not operated on; wounds healed. *Jan. 26th,* 1863.—Showed himself to-day; the truss has shifted, and fails to keep up the large rupture when at work; there is also some bulging of the side operated on, but he does not feel any of the pain complained of before the operation. Truss to be readjusted to the parts, and constantly worn.

Case 59.—Captain G——, æt. 42, operated on Nov. 14th, 1862, for an oblique inguinal hernia of the left side, of some years' standing. It is of considerable size, scrotal, and incapacitates him for active duty. The ordinary wire method employed. Had no pain after the operation, and slept at night without sedative; pulse and tongue unaffected.

17th.—Some œdema of the scrotum, and enlargement of the testis, but no pain or tenderness; no discharge; bandage removed. *19th.*—Bowels opened by castor-oil; no pain; appetite better; complains of a little bilious headache. *24th.*—More discharge; consolidation very evident in the canal, and in the fundus of the sac. *Dec. 2nd.*—Going on without an unpleasant symptom; pressure applied by a bandage over the wire loop. *6th.*—Has continued the pressure up to the present time, with the effect of a little increasing the size of the testis. Wire withdrawn with ease; granulations apparent at the punctures; consolidation in the canal very evident. *10th.*—Walked about his room with a pad and spica bandage. *14th.*—Truss applied. *20th.*—Left to-day for the country. The external ring closed; some consolidation still remaining in the canal, and a slight bulge, on coughing, at the site of the deep ring. Will let me know of any change.

Case 60.—F. H——, æt. 40, operated on Dec. 22nd, 1862, for a right, oblique rupture of ten months' standing, of the size of a small orange. At the bottom of the sac could be felt a roundish, hard mass, apparently a portion of consolidated omentum adherent to the fundus. No truss could be worn on account of the pain felt in this part, and all other apparatus ineffective in keeping up the rupture. Operated on by the wire method, care being taken to force up into the deep opening of the canal the adherent omentum, and to apply the wires outside the sac. Although the patient had been previously in delicate health from dyspeptic symptoms no bad symptoms whatever followed the operation. The tongue remained clean, and the pulse unaffected, except an occasional intermission for a day or two, the result, apparently, of the chloroform. On Jan. 14th, 1863, the wires were removed without difficulty, and water-dressing applied. On the 26th he left London on pressing business with the wounds nearly healed, and wearing Bourjeaud's apparatus. *March 1st.*—Have heard from this patient to say that the parts are firmly closed, with no tumour or cough impulse, and that he is wearing lever-spring horse-shoe truss pad.

Since the foregoing cases, I have operated on four cases in King's College Hospital—two boys and one girl, by the method with pins; and on a very large case of double scrotal hernia, in a man beyond the middle age, at his own urgent request, by the crossed wire operation. Three of the cases have been discharged cured; one, the girl, remains still in the hospital. No constitutional symptoms have arisen in any of these cases.

THE END.

www.ingramcontent.com/pod-product-compliance
Lightning Source LLC
Chambersburg PA
CBHW031852220426
43663CB00006B/593

LONDON:
SAVILL AND EDWARDS, PRINTERS, CHANDOS STREET,
COVENT GARDEN.